STEM Like a GIRL

For Nolan and Bennett,
my greatest experiments

STEM Like a GIRL

Empowering Knowledge and Confidence to Lead, Innovate, and Create

SARAH FOSTER
founder of STEM Like a Girl

BLACK DOG & LEVENTHAL PUBLISHERS
NEW YORK

Black Dog & Leventhal Publishers
Hachette Book Group
1290 Avenue of the Americas
New York, NY 10104

www.hachettebookgroup.com
www.blackdogandleventhal.com

First edition: September 2021

Black Dog & Leventhal Publishers is an imprint of Perseus Books, LLC, a subsidiary of Hachette Book Group, Inc. The Black Dog & Leventhal Publishers name and logo are trademarks of Hachette Book Group, Inc.

Additional copyright/credits information is on page 238.

Print book interior design by Katie Benezra

Library of Congress Cataloging-in-Publication Data
Names: Foster, Sarah, author.
Title: STEM like a girl : empowering knowledge and confidence to lead, innovate, and create / Sarah Foster.
Description: First edition. | New York, NY : Black Dog & Leventhal, 2021. | Audience: Ages 8–12 |
Summary: "STEM Like A Girl teaches important STEM concepts while showcasing dozens of curious girls showing off their scientific prowess. Filled with photographs of the girls and the experiments they've done, this guide will empower girls with knowledge and confidence to become future problem solvers and leaders in the scientific world and beyond" —Provided by publisher.
Identifiers: LCCN 2020044408 (print) | LCCN 2020044409 (ebook) | ISBN 9780762472604 (trade paperback) | ISBN 9780762472611 (library binding) |ISBN 9780762472598 (ebook)
Subjects: LCSH: Women in science—Juvenile literature. | Women scientists—Juvenile literature. | Science—Vocational guidance—Juvenile literature.
Classification: LCC Q130.F67 2021 (print) | LCC Q130 (ebook) | DDC 500.82—dc23
LC record available at https://lccn.loc.gov/2020044408
LC ebook record available at https://lccn.loc.gov/2020044409

ISBNs: 978-0-7624-7261-1 (hardcover); 978-0-7624-7260-4 (trade paperback); 978-0-7624-7259-8 (ebook)

Printed in China

1010

10 9 8 7 6 5 4 3 2 1

CONTENTS

A NOTE TO PARENTS

As parents or other adult caregivers, you play a key role in your daughter's developing interest and confidence in STEM. Studies show that young girls are interested in doing hands-on STEM activities but aren't given enough opportunity for this experience. Around middle school, many girls start to self-select out of advanced STEM classes because they lack both experience and confidence. This is why it is critical to engage them at a young age and continue cultivating their interests as they grow. Many parents want to foster a STEM identity for their daughters but are unsure how to do that or need resources to get started. This can be especially true if you aren't formally trained in a STEM field. Jo Boaler, a professor of mathematics education at Stanford University, has said, "We know that when mothers tell their daughters, 'I wasn't good at math in school,' their daughters' own achievement goes down." As parents, your words and actions, whether conscious or unconscious, have a dramatic effect on your daughter. With your support and encouragement for her to explore STEM activities, to try hard things, and to not be afraid of failure, your daughter will learn important critical-thinking skills and gain confidence. It can be overwhelming to plan and engage your daughter in STEM activities, which is why the activities in this book are intended to be easy to complete using supplies that you probably already have at home or that are readily available for a low cost at a local store or online. While the projects are designed for your daughter to complete independently, you are encouraged to work with her so you can learn and have fun together. Several of the activities are open ended, allowing you to get creative with your design and solution. You can work with your daughter as a team or challenge her to see what different ideas you each can come up with. No matter what, encouraging your daughter's interest in STEM will give her the support she needs and empower her to be successful in the future.

INTRODUCTION

For years, careers in science, technology, engineering, and math (STEM) were thought of as jobs only men were good at. Some people actually thought that the girl brain was not wired to do hard science or math problems! Recently, some progress has been made in shifting this thinking, but women are still underrepresented in almost all STEM fields. This mind-set doesn't start with adults. At a young age, many girls convince themselves that they aren't good in math or science. Sometimes this is because we hear it from a teacher, classmate, or family member. Sometimes it's a thought inside our own heads. Unfortunately, sometimes it's suggested to you only based on the fact that you are a girl. But girls are just as capable of doing STEM as boys. Doing something "like a girl" is NOT a negative thing! Whether it's running, throwing, or learning and experimenting in STEM, doing things like a girl should empower you. Girls have so much to offer in STEM, but often they don't really know what STEM is or all the amazing things scientists and engineers do. This book will teach you all about what STEM actually means and why girls should even care about it. (Hint: Doing STEM activities now will build confidence, creativity, and problem-solving skills you will need in every aspect of your life—and it's really fun!) You will see girls just like you who are doing STEM activities in their everyday lives and how this makes them feel strong and bold. And you will get to experiment and design like an actual scientist or engineer with fifteen STEM activities you can do right now at home. You are the future of STEM, and it's time to show the world what it means to STEM like a girl!

What Does STEM Stand For?

Have you ever looked at your toaster and wondered how those little red coils heat up your bagel in the morning? Or taken medicine when you're sick and wondered how it knows what to do inside your body? These are things scientists and engineers ask themselves every day as they study STEM. STEM stands for science, technology, engineering, and math. But what exactly does that mean? STEM is a way of thinking about solving problems that affect our community and everyday lives. Our world is constantly changing and shifting as new knowledge or things are discovered or new problems arise. Instead of thinking of science, technology, engineering, and math as four separate fields, STEM recognizes that these concepts overlap and work together.

Before we can understand why these fields are grouped together as STEM, let's first look at each of them separately to get an idea of what they mean, how they are different, and what someone who studies them actually does.

SCIENCE

Science is about observing and studying natural things in the world around us. From tiny atoms to massive galaxies, plants to animals, rocks to oceans, science is about figuring out how our natural world works. If that seems like a lot, that's because it is! Science is often broken down into different fields of study based on what the focus is. It would take way too long to list all the different types of science here; there are more than fifty different fields! However, you can think about the different fields of science based on what they study: living versus nonliving things. *Life science* covers all the living things in our universe, and *physical science* looks at anything that's not alive. That still might sound pretty broad, so let's break it down a little bit more.

As you can probably guess from the name, life science is the study of life and all living organisms, including how they work and interact with each other and their surroundings. Most people think of biology as the core of life science, but even it can be divided into over ten different categories based on what is being studied (again, too many to list here). Basically, anything that's alive is being studied and observed by some type of life scientist. Life science includes understanding the individual cells in our bodies (cell biology), how those cells work together to form our organs (physiology), and how our bodies work as a whole (medicine). Life scientists study bacteria (microbiology), animals (zoology), insects (entomology), and plants (botany), and how all those things live together and interact in our environment (ecology).

Studying how all living things work and interact with each other is important for many reasons. For example, understanding how viruses and bacteria infect both plant and animal cells helps scientists develop vaccines and antibiotics to keep humans, animals, and plants healthy. Or by learning what insects are either beneficial or harmful to crops, scientists can help protect our farms and agriculture, which are vital to our food supply.

So, if life science covers all living things, physical science covers everything else in our universe that isn't alive. Chemistry, physics, and earth science are often thought of as the main components of physical science. They build on each other, from the smallest subatomic particles to the massive solar systems and galaxies that make up the universe we live in. Chemistry looks at the study of *matter*, which is the scientific word for what things are made of. Think of chemistry as the study of building blocks of the universe. Chemists study how these different building blocks, or matter, interact and change under different conditions. From these key building blocks comes the field of physics. Physicists study how that matter moves through space and time and the energy and force it creates. Earth science is the study of our Earth and the space around it.

Observing and understanding the nonliving building blocks that make up our world and how this matter behaves and interacts is important for many reasons. Using the principles of chemistry and physics, scientists can create new fuel sources that are more efficient and cleaner for our environment. Earth scientists can help predict environmental changes and determine ways to protect our planet. They can understand weather patterns and are able to warn people of potentially dangerous hurricanes, tornados, or earthquakes.

TECHNOLOGY

When they hear the word "technology," a lot of people only think about computers and coding. But technology is so much more than that. Technology is a broad term that means using tools, materials, and machines to solve real-world problems and create useful or helpful inventions. Technology means applying what we learn from STEM to make actual systems or products. An example where all these fields intersect would be the development of an improved smartphone. This technology combines science, engineering, and math in many different ways. For example, science is used to develop chemicals that will provide a longer battery life for the phone. Engineering is used to design stronger materials that can withstand being dropped and getting wet. And math is needed to figure out how to make the processor chips run faster and improve the equations for facial-recognition capabilities.

It can help to think of technology as more of an idea than a specific field. It describes the innovation and process that draws ideas from all areas of STEM to make technological advances in our society. Science, engineering, and math help us learn and understand how our world works, while technology is what takes those observations, calculations, and designs and turns them into something that is actually helpful in our world. You can think of technology as a bridge. On one side of the bridge is a question or unknown. On the other side is the solution. Technology is the bridge that runs from the question to the solution and connects everything together.

ENGINEERING

Engineering involves applying science (and other parts of STEM) to design, build, and improve the machines and physical structures all around us that

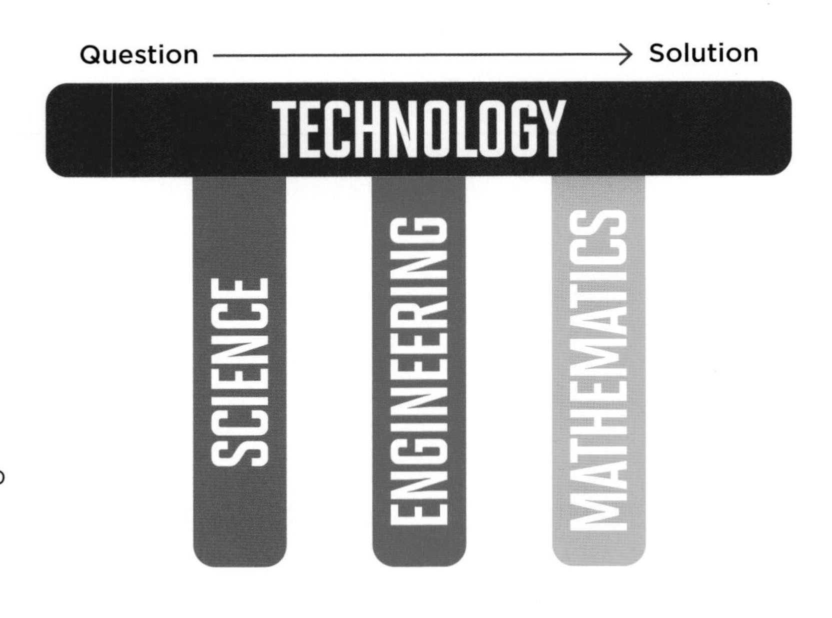

we use every day. Just like science, engineering is broken down into different branches based on what is being worked on and what problems need to be solved. There are four main branches of engineering: *chemical*, *civil*, *electrical*, and *mechanical*. Each of these can be broken down even further (again, there are way too many subcategories to list here).

Chemical engineers figure out the processes that turn chemistry (and actually biology, physics, and math too) into things like medicine, fuel reactors, and food products. They design ways of using materials like plastics, ceramics, or metals to make medical devices like artificial organs or cars that keep us safer if we are in an accident. Chemical engineers also work on making existing processes and procedures more efficient so they use less energy, create less waste, or can be done faster and more safely.

Civil engineers design the structures we live in and travel on every day. Buildings, roads, bridges, airports, and systems for our water supply are all part of what a civil engineer may work on. Civil engineers help to improve and protect the environment by designing ways to treat rainwater runoff to prevent pollution of our rivers and oceans. They design buildings and roads that will withstand earthquakes or other natural disasters. Civil engineers also plan neighborhoods and city parks for you to live and play in.

Electrical engineers work on the circuits and electronics of anything that uses power. They use the physics and math of electricity and electromagnetism to make all sorts of electronic devices that we need to power our lives every day. Electrical engineers may work on methods to make tiny circuits for a cell phone, sensors and controls for robotic machines, medical equipment such as CT or MRI scanners, and renewable energy sources.

Mechanical engineers design any machine or system that needs motion to operate. Mechanical engineering combines principles from math and physics to make lots of different equipment that has moving parts. They build engines for cars and airplanes, machines that are used in all sorts of manufacturing plants, turbines for wind energy, and even the motor on a toy robot. Whether something has just one moving part or a whole system of moving parts, if it moves at all, you can be pretty sure a mechanical engineer had something to do with it!

MATH

Math is all around us, in STEM and in the world! From counting out money when you buy something at the store to measuring ingredients to make cookies, math is something you need and use pretty much every day. Mathematics is the study of numbers, shapes, and patterns. People often find math scary because they don't understand how all those equations and formulas apply to real life. But chances are, you are using math even when you don't realize it. Math helps you figure out how long it will take to drive to your friend's house so you arrive on time. It can help you design the layout of your bedroom so you know where your furniture will fit. Math is even used in fashion design for measuring clothing and piecing fabric together in the right pattern.

Scientists and engineers use math in many different ways. For example, geometry is the study of shapes and patterns. A civil engineer uses geometry for designing and building structures. Geometry is also used by computer programmers for drawing the animations in movies and video games. Calculus allows us to understand how things change over time. Calculus helps physicists explain why the planets move the way they do in their orbits and helps environmental and conservation scientists analyze how populations of predators and prey change over time. This helps them predict if a certain species is at risk of becoming extinct. Biologists use calculus to figure out the growth rate of bacteria or other cells, and how different conditions such as temperature affect their growth.

On the surface, math looks like a complicated bunch of numbers, letters, and equations, but when you know how math applies to real-world problems, it becomes a powerful tool for use in STEM. Stephanie Salomone, chair of the mathematics department at University of Portland, has said, "Mathematics is not just about calculations or really long formulas or solving for x. It's about finding patterns, thinking critically, and solving problems that address both local and global issues." When you think of math in this way, maybe those hard equations don't seem quite so scary.

4X

Asking Questions, Solving Problems

So now you know what each of the letters in "STEM" stands for, and you have a good idea of what those fields actually involve. But why do we group them together as STEM? Instead of thinking of science, technology, engineering, and math as four separate fields, STEM recognizes that these concepts overlap and need to be considered together. Understanding the science of why something happens isn't helpful without having the technology and engineering to apply that information in order to build something useful or solve a problem. And nothing in STEM would be possible without the math to accurately calculate each process or system. STEM allows us to address each piece of a problem while working toward the whole solution.

STEM is all around us and impacts our lives every day. It's easy to think that STEM is just for adults and that kids are too young to be scientists or engineers. But learning about STEM early on is so important, and it's something you might not even realize you are already doing. Think about a toddler playing on the floor with blocks. Each time they stack their blocks and knock them over, they are learning about building structures, stability, and cause and effect. They are being engineers! When you bake cookies, you are using chemistry by mixing together different ingredients that interact to form the perfect soft and chewy cookie. Or even just by watching a bee in your backyard flying from flower to flower, pollinating as it goes, you are exploring environmental and conservation science and using your power of observation, a critical piece of STEM. Jane Goodall, a famous scientist and conservationist, said, "You cannot get through a single day without having an impact on the world around you. What you do makes a difference and you have to decide what kind of difference you want to make." Each time you make an observation or try to figure something out, you are being a scientist or engineer.

You cannot get through a single day without having an impact on the world around you. What you do makes a difference and you have to decide what kind of difference you want to make.

—JANE GOODALL, SCIENTIST AND CONSERVATIONIST

One thing all scientists, technologists, engineers, and mathematicians have in common is that they ask A LOT of questions. Kids are naturally curious and ask a lot of questions, so you are probably already acting like a scientist or engineer. STEM involves looking at an existing problem and asking what the solution could be. Or maybe a solution already exists, but it needs to be smaller, faster, or more efficient. Scientists and engineers are always asking questions like why does this work? What is the solution? How can this be improved?

> The best scientists and explorers have the attributes of kids! They ask questions and have a sense of wonder. They have curiosity. "Who, what, where, why, when, and how!" They never stop asking questions, and I never stop asking questions, just like a five-year-old.
>
> —SYLVIA EARLE, MARINE BIOLOGIST

Asking questions is the best way to dig deeper into a problem, yet many people (kids and grown-ups alike) are often too embarrassed to ask questions. Sometimes we are afraid it will make us look dumb. Or that we are the only people in the room who don't know the answer. But STEM encourages asking as many questions as possible, because that's what promotes curiosity and exploring the unknown. Many of the great inventors we think of today had the courage to ask questions, and some were even made fun of for asking those questions, but that didn't stop them! Albert Einstein was well known for his ability to ask questions and remain curious. He said, "The important thing is not to stop questioning. Curiosity has its own reason for existing." If Einstein had the courage to ask questions, so should you!

With STEM affecting so many different areas of our lives, teamwork is also really important. STEM brings people from different backgrounds together in teams to address all the parts of a problem. The team works together to come up with a well-rounded solution. People often picture scientists and engineers working alone in a lab, hovering over an experiment. Or maybe you picture them typing away at a computer by themselves for hours on end. But for most people in STEM, that's not usually the case. Scientists and engineers know that cooperation and communication are two critical skills for problem-solving. STEM is about combining each person's talents and expertise to solve problems. This helps them come up with solutions that they might not have thought of or imagined on their own. It allows each member of the team to contribute in a meaningful and helpful way.

Why Should Girls Care About STEM?

By now you should be convinced that STEM is impossible to avoid and is a huge part of your everyday life. But why is it so important to get *girls* interested in STEM? A lot of girls think that either you are born being "good" in math and science or you aren't. Maybe you've even heard a friend or your mom say, "Science is just too hard! I could never do it." If those skills don't come easily to you at first, it's tempting to give up and follow others by saying to yourself, "I'm just not good at math." But just like anything in life, STEM involves skills you need to work hard at and practice. No artist is born knowing exactly how to paint or sculpt. No ballerina is born knowing how to do the perfect plié or arabesque. They work hard for years and practice over and over to become great. STEM is no different. No one is born knowing how to do math or how to build a chemical reactor. Like any muscle, the more you practice STEM activities, the more you learn.

We look at science as something very elite, which only a few people can learn. That's just not true. You just have to start early and give kids a foundation.

—MAE JEMISON, THE FIRST AFRICAN AMERICAN WOMAN TO FLY IN SPACE

Sometimes girls think that if they like fashion, art, and dance, they can't also be interested in STEM. Since when do we have to choose one over the other? Girls are not limited to only a few interests. There is so much creativity, design, and art that goes into STEM. It's okay to have an interest in fashion and also like chemistry. Maybe you'll design the next moisture-wicking fabric made from sustainable materials. Or maybe you are really into music. Did you know that the sound we hear is actually vibrational energy? Don't think that just because you are creative or like "girly" things, you can't also be into STEM.

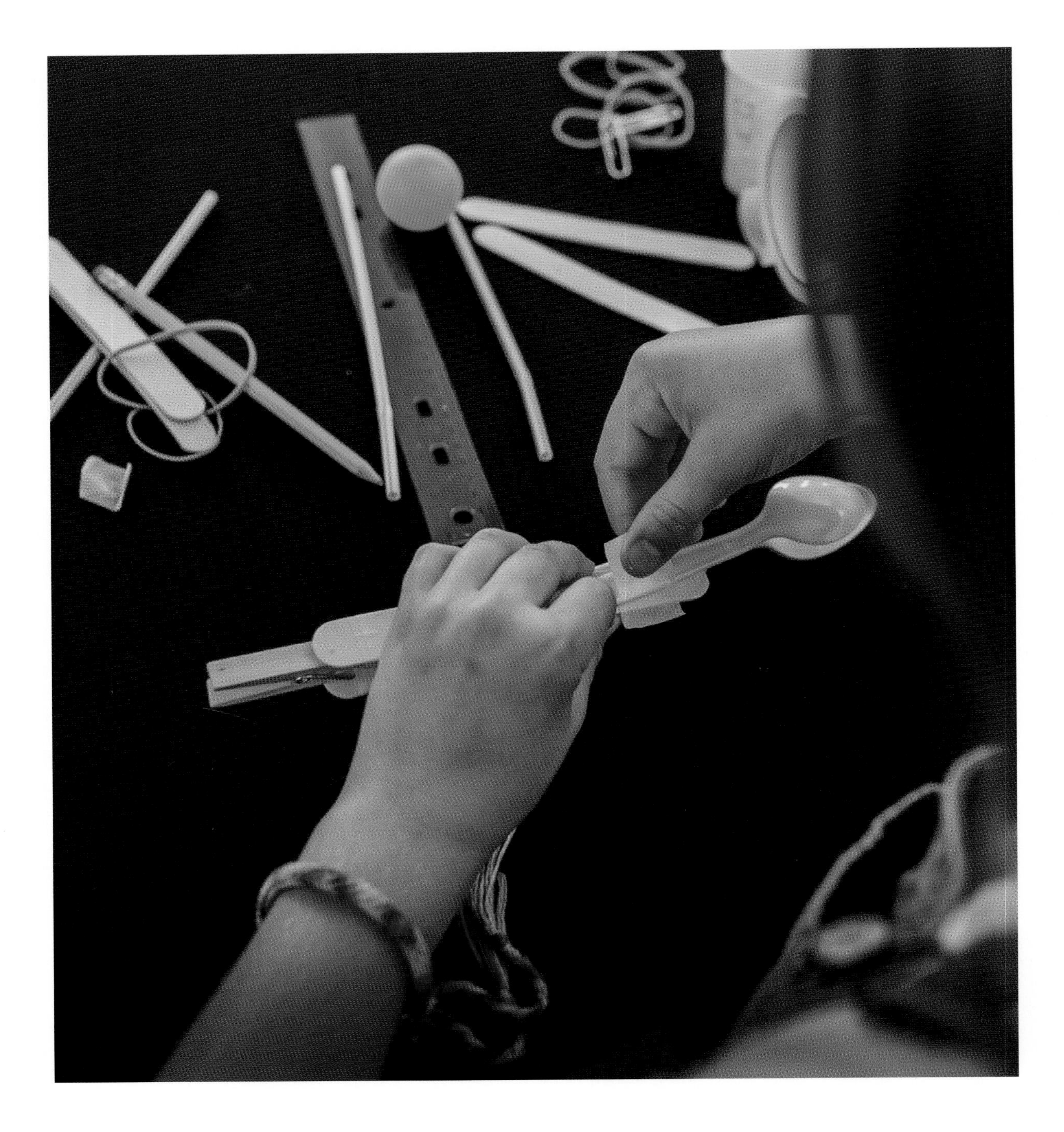

It's okay to be a princess, but I want girls to build their own castles too.

—DEBBIE STERLING, ENGINEER AND FOUNDER OF GOLDIEBLOX

One of the biggest reasons that you should care about STEM is that it helps prepare you for pretty much everything in life. As you engage with the world around you, you will inevitably come up against problems or situations that you don't know the answers to. Taking part in STEM activities now helps to develop your critical-thinking skills and build your self-confidence. Sometimes problems feel so huge and impossible that it's hard to figure out where to even start. STEM teaches you how to ask the right questions, brainstorm possible solutions, and evaluate your ideas. Each time you let yourself take a risk and try hard things, you give yourself the opportunity to learn something new. STEM offers us opportunity after opportunity to try new things, take chances, and put yourself out there.

STEM also gives you the tools to become a strong leader in your school and your community. STEM teaches you how to make observations about the world around you and identify problems or areas for improvement. A good leader knows how to brainstorm ideas, test solutions, and work with others to solve a problem. You may have to work in a group on a project at school. Good leadership skills will help you guide your team to work together, communicate, and solve the challenge. Later in life, at work, you may have to lead your team in troubleshooting a problem that comes up in your business. The critical-thinking skills you learn from doing STEM activities now will teach you to be a strong leader and help you navigate even non-STEM problems.

Studies find that many girls are drawn to activities and careers that help others. When asked what is important in a career, girls and women often use phrases like "making the world a better place," "caring for others," "making a difference in someone's life," or "working toward a common goal." In school, science and math are often taught in a very technical way, and we don't see how those skills can apply to real-world problems that interest us. But STEM provides so many exciting and rewarding career opportunities to make a difference in the lives of other people.

House
House
Apartment
House
House

Civil engineers work to design and build more accessible parks and playgrounds so children with physical differences can enjoy them. Biologists and chemists research new medicines that will help save people's lives, cure illnesses, or help people live more comfortably with a disease that we haven't yet found a cure for. Environmental scientists and conservationists study how climate change affects plants, animals, and humans. They work to protect our environment to ensure that future generations have a safe planet to live on or prevent a species from going extinct. If STEM is something that interests you, it's a perfect field for helping others and making a difference in the world. Through STEM, you have the opportunity to shape our future!

Girls and women also bring a different perspective to STEM and solving problems. Not only are women more than capable of working in STEM; we also bring knowledge, experience, and skills that men don't have. Since half the population is female, it is important that women are involved in the design process so that our needs and perspectives are also considered.

Women are not just smaller versions of men. Our bodies are different, and we have different needs. These differences needed to be considered, for example, in the design of airbags and other crash safety features in cars. For years, crash tests were performed using only male-sized dummies. It wasn't until 2011 that companies started testing cars using smaller dummies that better represented a female body. When they started testing on smaller dummies, they found an increased risk of death for the smaller passengers. If women had been involved in the design process and made their voices heard earlier, maybe those issues would have been addressed sooner. There are also many diseases that mainly affect women, including breast cancer, ovarian cancer, and certain medical conditions that can lead to infertility. Having women in STEM working on treatments for these types of illnesses brings a more personal perspective to the problem, since it's something that directly impacts us.

Besides having physical differences, women also add a unique way of thinking that brings diversity to a team. Research has shown that the brains of men and women are wired differently, allowing each gender to think and process information in distinct ways. This means that men and women each offer strengths that can be applied to STEM and problem-solving. It doesn't mean that one way of thinking is better than the other but that both perspectives are needed to complement each other. Generally speaking, women tend to think in broader terms and focus more on the big picture before jumping into problem-solving mode. Women also tend to have heightened language and communication skills, which help them talk through situations and conflicts. Additionally, women often have a greater sense of empathy toward others, allowing them to feel more connected with the people they are trying to help. Including both men and women on STEM teams leads to increased diversity and perspective when approaching a problem and determining possible solutions.

There are so many reasons for girls to get involved in STEM, both now and in the future. Not only does learning about STEM give you valuable tools to be successful in life, but you have so much to give to STEM. The world needs to hear your voice, so don't be afraid to share it!

Failure Is a Good Thing!

Do you like to fail? Most people would probably answer *NO WAY!* Think about a time you tried to do something and it didn't work out the way you thought it would. How did you feel? You may have felt like you wanted to give up because the task seemed too hard. Maybe you felt angry or embarrassed. Failure can make us feel frustrated, defeated, or like we aren't good enough. It can feel even worse when you see friends, classmates, or people you look up to who seem to be succeeding and wonder what you are doing wrong.

But failure is a part of life, especially in STEM. Whether it's learning to walk, learning to ride a bike, or inventing the lightbulb, we generally have to fail before we succeed. What would happen if everyone stopped trying the first time they failed? Chemist Stephanie Kwolek discovered the bulletproof material known as Kevlar after testing a substance that most of her fellow chemists would have thrown out. It didn't fit the properties they were looking for, but she took a chance and invented something that has saved many lives.

Behavioral neuroscientist Elizabeth Gould was told she had made a mistake in her research when she discovered that some adult brain cells (called *neurons*) could heal themselves and regrow. Instead of listening to her critics and giving up, she kept at it and after years of research proved she was correct. Her data was so conclusive that the original scientist who said she was wrong (a man!) finally admitted that neurogenesis, the growth of new neurons, was real.

So, what if we thought about failure as a good thing instead of something negative? Often we can learn more from our failures than our successes. Think back again to that time you tried something hard and it didn't work. After your initial feelings of frustration, if you look closely, are there things you learned from your attempt? When you make observations about how or why things don't work or "fail," you can gain valuable information that can improve your idea or design. Think of the word "fail" as an acronym for "first attempt in learning." Each time you try something new or hard and it doesn't work, ask yourself what you can learn from it. Sometimes failed attempts end up being the best teachers.

When you try hard things, failure isn't just okay; it's pretty much a given. It's unreasonable to think that every design, every experiment, or every idea is going to work out right the first time. But trying something hard and failing doesn't make *you* a failure. It makes you courageous. It makes you resilient. It proves to yourself and others that you are strong and willing to take chances. To step outside your comfort zone. It's how we handle the failures and how we allow them to shape our next steps that determine our ultimate success. Don't let your fear of failure stop you from trying, because you never know what you may learn or accomplish along the way. Where will your failure take you?

F FIRST
A ATTEMPT
I IN
L LEARNING

The Design Process

We've talked a lot about how STEM involves problem-solving and critical thinking. But how exactly do scientists and engineers accomplish this? When doing a STEM activity, it can be really tempting to jump in and start designing or experimenting right away. Occasionally this works, but more often it leads to poorly thought-out results and frustration. Taking the time to work through a problem-solving method not only leads to better results but can also save time. Scientists and engineers often use what is called the *design process* when approaching a challenge. The design process follows a series of steps that guide teams through the problem-solving process. The design process is especially helpful when you have an open-ended challenge because it actually encourages failure through trial and error. Remember, failure can be a good thing. It's how we learn what doesn't work so we can make improvements.

The first step in the design process is to *ask questions*. In STEM, it is important to ask a lot of questions so that you are clear about what you are being asked to do. These questions may include: What is the problem you are trying to solve? Who are you designing this project for, and what are their needs? What are the requirements, and what are the limitations for the design? What is the end goal for this design? This first step is critical, especially when working in a team. It makes sure that everyone is on the same page and that you know exactly what you are trying to accomplish with your design.

The second step is to *imagine and brainstorm ideas*. This step might include doing research about the problem you are trying to solve. Try to learn as much as you can using all the resources available to you. You might want to talk to experts in the field, people with different STEM backgrounds, or the end user of the design. Try to come up with several ideas or possible solutions, because often your first thought doesn't turn out to be the best one. This is where you can really get creative and think of lots of different options. If you're working in a team, be sure to listen to each other's ideas and work together to come up with the best solution. Build on each other's ideas. Don't rule anything out, even if an idea sounds crazy at first. By the end of this step, you should have

several ideas to consider when moving forward in the design process.

Once you've asked lots of questions and imagined several possible solutions, it's time to choose your best idea. This is the *planning* step. Spend time going over your different design ideas and pick the one you want to try first. If you're working in a team, this can sometimes be the hardest step, especially if people disagree on what to try first. Part of planning is to draw out your design and gather all the materials you will need. Here you should also think about how you will test different parts of your design to see if things are working.

Experiment Test Sheet
Experiment Name:
Date:
Test Number
Observations/Thoughts
Conclusions:

Now it's finally time to start *creating*! Hopefully, all the prep work you put into the design process will pay off as you put your plan into action. This is the fun part as you see your ideas come to life. As you build and test your initial design, you might find that you need to make changes, and that's okay. You'll find that often things don't work out exactly as you thought they would. Maybe a certain material isn't strong enough to support your design. Or one of the parts doesn't move the way you expected it to. Observe and evaluate what worked well and what didn't.

Using the information you gained from the initial design and testing, determine ways you can *improve* your design. This is where you go back and make changes based on your observations. You may have noticed that the design process is circular. It doesn't have a set beginning and end. Take the information you learned from testing, and make changes to your design. Sometimes you may only need to make small adjustments. Other times, your initial design may totally miss the mark. In these cases, it might be best to start back at the initial "ask" step or see if you overlooked something critical in your planning.

Remember that all failure, whether big or small, is an opportunity to learn and improve. By working through the steps of the design process, you can apply your creativity in a well-thought-out way and be confident in your end result.

How to Use This Book

So, what can girls your age do right now to learn about STEM? You're off to a great start by reading this book! You'll read interviews with girls just like you who are doing STEM in their own unique ways. These girls share their stories of how they first became interested in STEM. They talk about not only their successes in STEM but also about their feelings and struggles when a design or experiment doesn't go how they thought it would. They share how STEM helps them feel brave and confident in all parts of their lives. The girls also talk about lots of ways they do STEM at home, in their schools, or with different groups. Sometimes it can feel as though you are the only girl in your class who is interested in science or math, but seeing other girls doing really cool stuff in STEM can help you feel like part of a bigger group. Their stories will inspire you to take chances and try hard things.

The best way to actually learn STEM is by simply jumping in and doing it. At-home STEM activities are a great way to explore and learn about different topics whenever you want, since you don't need to go to a class or to the science museum. You will learn how to plan out an experiment, test your design, and make observations. Plus, it's a lot of fun! This book offers fifteen different activities that can be done with supplies you probably already have at home or that are readily available at a local store or online for a low cost. At the end of each activity, there's a section called "STEM Application" that explains the STEM concept and real-life relevance of each activity.

All the activities have step-by-step instructions and pictures to guide you. As you work through the steps and experiment, make observations and ask questions about what you are doing. Feel free to get creative, try different things, or change your design. Remember, STEM is all about curiosity and trying new ideas. You can make your own experiment test sheet like the one shown on page 37 to record your work and document your experiments.

The last four activities, starting on page 192, are open-ended projects. This means that there is more than one way to solve the problem. These activities are called "design challenges." You will notice that there are fewer instructions and pictures showing you exactly what to do. Instead, you get to use your creativity and problem-solving skills to think like a scientist or engineer and design your own solution. There is no right or wrong way to complete these challenges!

Like many things, STEM is more fun with friends. Scientists and engineers often work in teams to solve problems, so you are encouraged to do the same. As you work on the activities, grab a friend or family member, and have fun experimenting together. Challenge each other to see what different designs each person can create, or work together as a team. Spend some time sharing your ideas or design when you are done. Remember that each person might have different ideas or strategies for solving the problem. One of the cool things about STEM is there often isn't just one right answer!

EXPERIMENT TEST SHEET

Experiment Name: ______________________________

Date: ____________

TEST NUMBER	OBSERVATIONS/THOUGHTS	SUGGESTED DESIGN IMPROVEMENTS

Conclusion: ______________________________

____

____

____

____

____

____

AT-HOME STEM KIT

Sometimes the thought of doing STEM activities at home can be overwhelming and intimidating (especially to the adults in your house!). To make doing STEM activities at home even easier, consider putting together a "STEM Essentials" kit. Keep this kit stocked and ready to go for when you want to experiment. The materials listed below are used in several of the activities described in this book, so have them handy. Some suggestions for your kit include:

- Straws
- Craft sticks of different sizes
- Rubber bands of different sizes
- Clear tape
- Masking tape
- Plastic cups of various sizes
- Construction paper or card stock
- Pipe cleaners
- Pipettes or droppers
- Scissors
- Pencils
- Markers
- Ruler
- Notebook
- Safety glasses

SAFETY NOTES

- Before starting any STEM activity, make sure to ask an adult in your house for permission first.
- Take some time to read through all the instruction steps for the project before you start. If you have any questions about a step, talk it over with your adult first.
- Some of the activities use mild chemicals or tools. It is important to protect your eyes by wearing safety glasses when working on these projects. These particular activities include a reminder to put them on before you start.
- When working with scissors or other tools, be careful. Always ask an adult for help if you aren't comfortable with a step.

THE GIRLS AND THEIR PROJECTS

Hazel, age 10

What are some things you want us to know about you?
I love art, animals, and cooking. I'm a very social person with big dreams. I don't like to be put in boxes, but if I were a cat, trust me, that would be my favorite activity!

Where do you do STEM?
I mostly do STEM at home because there's not usually too many opportunities at school. I guess I have really amazing parents who let me do amazing activities.

When did you first become interested in STEM?
I really started loving STEM when I was in third grade. My teacher was very encouraging, and she had a degree in botany. After that I felt more attached to STEM, and now it's become a big part of my life.

Why do you like doing STEM activities?
I've never been good at answering these types of questions, but I guess it's just really fun to play around with a soldering iron or study different varieties of wild berries. That's why!

What is your favorite STEM project you've ever done?
My dad makes sound pedals [electronic devices that change the sound of music], and we decided to make a distortion pedal together. It was pretty hard, but I got to solder and plan out a circuit board with my dad. It was really fun and I'm glad I did it!

When an experiment or design doesn't turn out how you expected, how does that make you feel?
Usually if I have enough supplies, I'll just try again, but if it's something like a pedal, I'll troubleshoot.

What other activities make you feel courageous, confident, and bold?
I'm a person who is very drawn to the stage, and I'll take any opportunity to sing in front of my school. I really like to make cool earrings, and when people compliment them, it makes me feel really good about myself.

Who do you look up to?
Honestly, I look up to my mom because in her everyday life she faces a lot of problems and that's what makes her amazing. She's always working hard on accomplishing her dreams.

What are some ways girls can do STEM activities at home?
You could make elephant toothpaste. It causes a fun explosion, and it's very easy to make.

What advice do you have for other girls?
Be strong! Society's picture of girls is long hair, skinny, and "perfect." It doesn't matter what race you are or your body type, you are AMAZING!

Daphne, age 10

What are some things you want us to know about you?
I go to a French immersion school and I speak French. I actually skipped third grade, so I'm now going into sixth grade. I love theater and was in a play that was a mishmash of a lot of the books from my state's Battle of the Books competition. I played one of the leads who is obsessed with musical theater, which is pretty much me! For the past two years, I've been part of a traveling theater group that goes around town and performs at retirement homes and our local children's museum. I also like to play soccer, basketball, and tennis. I do a lot of dance, and my favorite is tap. I also take piano and voice classes. I love reading and will read absolutely anything.

Where do you do STEM?
I do STEM at school and online at Girls Count workshops. I'm also making my own website at home.

When did you first become interested in STEM?
I've always loved math, and it's been my favorite subject. Ever since I was little, whenever anyone would ask me what I wanted to be, I've always said that I wanted to be an inventor.

Why do you like doing STEM activities?
I just like math a lot. I like how you can solve a lot of problems in the world using math. I'm pretty good at it and am usually the one to finish first when we do math problems.

What is your favorite STEM project you've ever done?
I'm currently working on a website. It started as a school project for science class about finding an ecological problem and solving it. My plan is to create a website that raises money to help reduce plastic pollution from entering the ocean. It's like a "choose your own adventure" game about a kid that saves the ocean from plastic pollution. I'm making a database of organizations to donate to based on what path you choose. If you make a wrong choice and don't save the oceans, the ending turns out sad.

When an experiment or design doesn't turn out how you expected, how does that make you feel?
At first, I get really mad, but then it makes it kind of fun. When I'm with a friend, we'll test and make changes, so it makes the whole experience more fun. After I get over my anger!

What other activities make you feel courageous, confident, and bold?
Theater, 100 percent. It makes me feel really proud, especially when I'm on the elevated stage. I feel like I'm looking down on the world and giving them a huge performance. It makes me feel really brave and strong inside, like, "Yes, I am doing this!" I put in so many hours memorizing songs, lines, and dance moves, and just to let it out and do it for real feels

really amazing. Also, because I skipped third grade, people often underestimated me until I won the spelling bee, which made me feel really amazing.

Who do you look up to?
My mom and my grandma. In STEM, I look up to Margaret Hamilton, who made most of the coding for the spaceship. Also, Ada Lovelace. She's the one who designed the first computer program ever. When she was only twelve years old, she even wrote a book called "Flyology" about her studies of birds, and she made designs for wings.

What are some ways girls can do STEM activities at home?
There's a lot of stuff online that you can do, even different workshops you can join when you want to take classes but can't go in person.

What advice do you have for other girls?
Dream big. Don't let anything get in the way of achieving your dreams. Be who you are, and don't give up!

Sydney, age 9

What are some things you want us to know about you?
I love unicorns. I think they are so magical, which is why I like them. I also like to play with Legos and to play Roblox on my iPad. I like to play with my cousin's dog named Panda.

Where do you do STEM?
I do STEM at school during class and when we do basic experiments. I also have a science kit at home from my sister.

When did you first become interested in STEM?
I saw that my sister brought home a robot that she built at a STEM class, and it was so cool. It could move, and I wanted to make something like that too!

Why do you like doing STEM activities?
The STEM activities are so creative, and I also learn!

What is your favorite STEM project you've ever done?
When my mom and I put an egg in a container and put stuffing in to keep it safe. I used fabric to wrap the egg. Then we dropped it from a high spot, and the egg did not break! I was nervous that it would break but also really excited. When I opened it up and saw that the egg didn't break, I was really happy.

When an experiment or design doesn't turn out how you expected, how does that make you feel?
I feel disappointed, but I know I should try again.

What other activities make you feel courageous, confident, and bold?
When I work together in a group on a project. Sharing my ideas with others makes me feel confident. I like having other people to talk to, and it makes it more fun.

Who do you look up to?
I look up to Cynthia Breazeal, who is a professor at the Massachusetts Institute of Technology. She makes robots that can understand expressions in humans and other stuff like that.

What are some ways girls can do STEM activities at home?
One way is you can cut out cardboard and tape Legos together to make a machine thing. Try making a roller coaster with your Legos to see how high you can build it or if you can make a cart travel on the tracks.

What advice do you have for other girls?
Have fun and try your best!

AIR CANNON DIFFICULTY LEVEL

Use a paper cup and a bag to make an air-powered cannon that propels a ball into a target.

MATERIALS

Paper cup
Pencil
Plastic fold-top sandwich bag or plastic
 produce bag
Masking tape or duct tape
Mini pom-poms (or mini marshmallows)
Target or bucket (optional)

METHOD

1. Using a pencil, poke a hole in the bottom of the paper cup, and rotate the pencil back and forth a bit to smooth out the edges of the hole. The hole should be about the size of the pom-pom but not bigger. Test the size of your hole by sticking a pom-pom into it. It should fit snugly but not push all the way through. If your hole is too small, use the pencil or your finger to make it a bit larger. If your hole is too big and the pom-pom pushes all the way though, start again with a new cup. Once your hole is the right size, set the pom-pom off to the side.

STEP 1

2. Securely tape a plastic bag around the open end (the top) of the cup to form a seal between the bag and the cup. The bag will be the power source for your cannon when it's filled with air. Use as much tape as you need to make sure all the edges of the bag are secured to the cup and no air can leak out.

3. Once you think the bag is secure, blow into the hole you made in the bottom of the cup to inflate the plastic bag. Look for any leaks where air could escape out of the bag or at the seal between the bag and the cup. If you find any, tape them shut so the air can't escape.

STEP 2

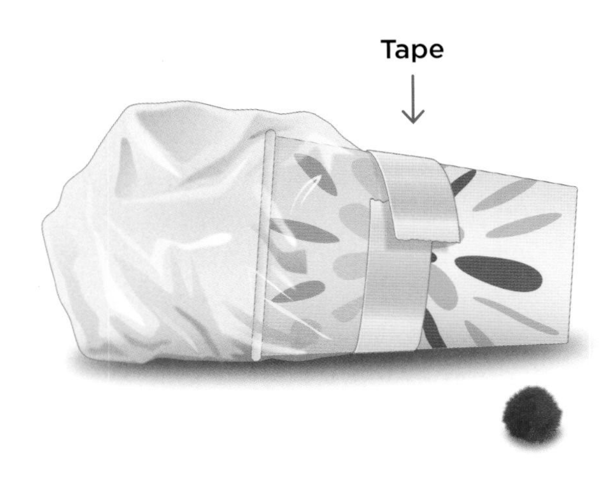

4. When you think you've completely sealed the bag to the cup, it's time to test your air cannon with a pom-pom. Blow into the hole so the bag inflates like a balloon. Work quickly but carefully to place the pom-pom snugly into the hole, trapping the air inside the bag and cup. Be careful not to let the pom-pom slip all the way through the hole.

5. Aim your air cannon away from people (or at the target or bucket, if using), and give the bag one firm squeeze to push the air through the cup and launch the pom-pom.

6. Make observations and record your results on your experiment test sheet. Did the pom-pom shoot out of the air cannon? Or did the air leak out, causing a failure to launch? What type of flight pattern did your pom-pom follow? Are you able to aim your air cannon so that the pom-pom hits your target?

7. Make necessary changes to your design, and test again.

STEP 4

STEP 5

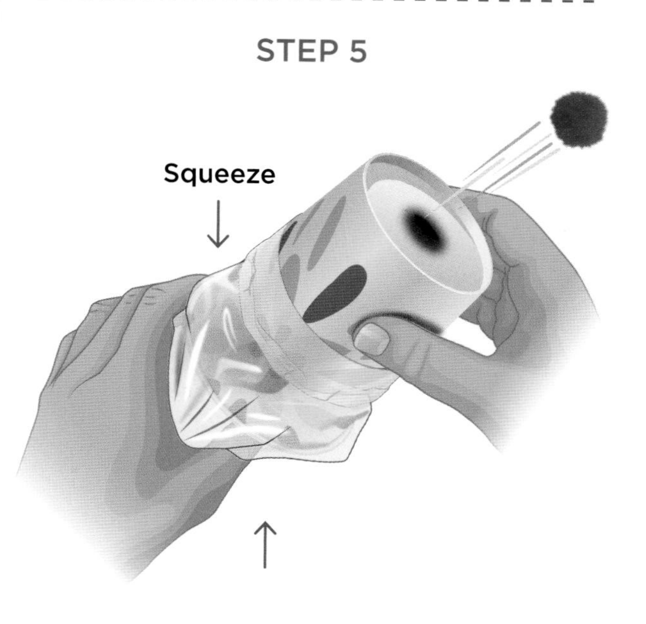

STEM APPLICATION

In order to make the pom-pom (your *projectile*) fire out of the air cannon, you need pressure behind it. When you blow up the plastic bag, you increase the air pressure inside the cup and bag. The pom-pom acts as a plug, trapping the air pressure inside the bag. When the bag is squeezed, you force the pressure to escape from the air cannon, causing the pom-pom to fire out. Did you have any air leaks in your air cannon while you were testing it? If so, you probably observed that the air escaped from the leak instead of pushing the pom-pom out of the cannon.

Isaac Newton's third law of motion states that for every action, there is an equal and opposite reaction. In this case, the action is squeezing the air-filled bag. The reaction is the pom-pom firing out of the cannon. The harder and faster you squeeze the bag, the farther your projectile will travel. *Trajectory* is the path the projectile follows. Trajectory is based on several things, including the angle of the cannon, the velocity or speed the projectile is moving, and the mass or weight of the projectile.

Áine, age 9

What are some things you want us to know about you?
I live in Portland, Oregon. When I grow up, I want to be an astronaut or work for NASA as a scientist.

Where do you do STEM?
I like to do STEM at school during STEM night, and we also do fun holiday-themed STEM projects during the year. I also do it at Girl Scouts, where I got to make a robot to earn my STEM badge.

When did you first become interested in STEM?
In pre-K we used to go to our middle school science lab to experiment. We also started doing coding in pre-K on our iPads. Something cool at our school is that we make science roller coasters in whatever theme we want.

Why do you like doing STEM activities?
I like it because you don't need to do just math or science; you can do both together, and that makes it really fun. It doesn't need to be just one thing. Basically, a lot of things are STEM. The projects are really fun, and you can make so many things, like robots, crafts, and number charts. Even baking needs math and science!

What is your favorite STEM project you've ever done?
I don't have just one! I have three favorites. One was when I got to do an egg drop at school out of the second-story window, and mine didn't break when testing! Another was when I got to make a model of the Hawthorne Bridge that's in my city. I also liked when I made a robot for Girl Scouts.

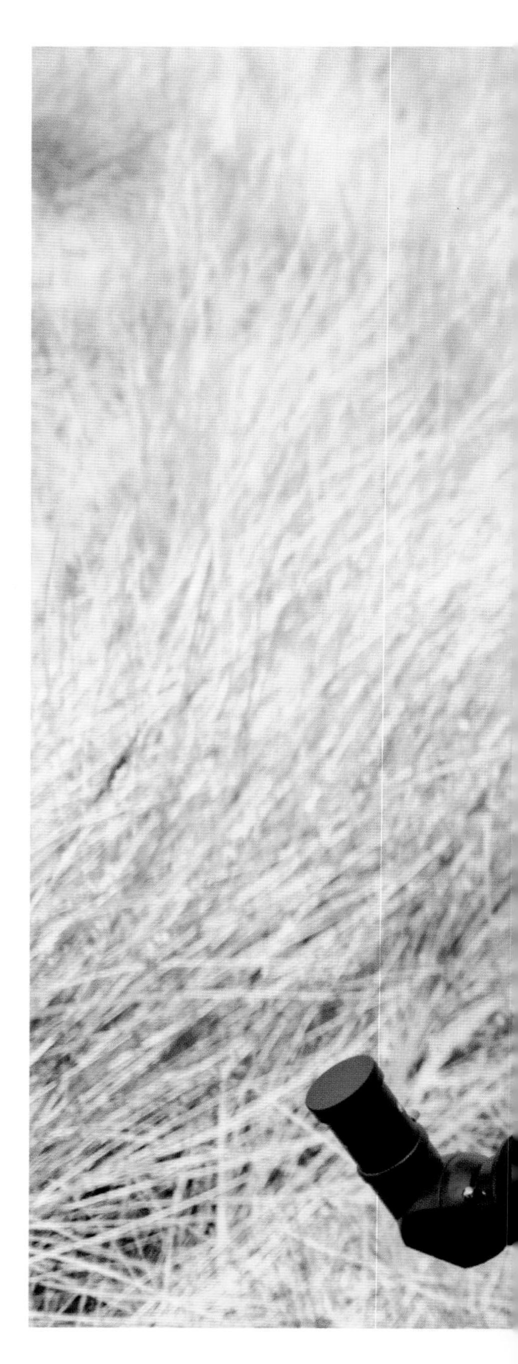

ROUTE 1: 1 Ursa Major 2 Ursa Minor 3 Boötes 4 Canes Venatici 5 Corona Borealis 6 Leo
ROUTE 2: 7 Orion 8 Canis Major 9 Canis Minor 10 Gemini 11 Taurus
POLARIS

When an experiment or design doesn't turn out how you expected, how does that make you feel?
I'm okay, and sometimes that's really good. You can make a different idea that's built into the same one. Like if you were planning to make something that's really cool and it doesn't work out, you can try again and add different ideas to your first idea to make it even cooler.

What other activities make you feel courageous, confident, and bold?
I really like doing art, and I like baking. I also like looking at the stars in space and trying to find some of the planets in our solar system. I saw Jupiter and Saturn once. I also saw the NEOWISE comet, which won't be back for another six thousand years. I've seen a shooting star and also the International Space Station.

Who do you look up to?
I look up to NASA employees and astronauts, because when I grow up, I want to work for NASA. I want to be an astronaut, and I really want to maybe be the first girl on the moon.

What are some ways girls can do STEM activities at home?
You can bake, because baking is basically science. You need to have the right amount of baking soda, salt, sugar, and flour. If you get it wrong, things can rise too much or shrink, and it turns out to be a big mess! Also, if you have scrap pieces of metal and batteries, you can try to make something that's robotic. You can twist things like hands out of bendable metal like pipe cleaners, and you can add batteries and try to make it work like a robot.

What advice do you have for other girls?
If you make a mistake on a project, don't give up. Just keep on doing it. It's actually really good to make a mistake because you can learn from it, and you can start over or build from it. If you start over, you can keep the idea and add on to it to make it even cooler. Since you know what happened to your experiment, you can think again to make it better. Just don't give up, and keep on doing what you do!

Carina, age 12

What are some things you want us to know about you?
I love playing soccer because it is a competitive sport. I have been playing since I was six years old. The best tournament ever was Soccer in the Sand at the beach. The coach allowed me to play at my older sister's tournament because I was good. I live on a small farm with lots of animals, including two dogs, barn cats, chickens, turkeys, pigeons, a pheasant, a horse, sheep, and goats. When I was younger, I would play soccer with my dog Chapo. My other dog, Borraow, chases me when I ride the quad around.

Where do you do STEM?
I do STEM at my library when they have events, and I have gone to different aviation camps through the Canby Eagles and also through Girls in Aviation. During the COVID-19 pandemic, I did a lot of STEM at home because things were closed.

When did you first become interested in STEM?
When I went to the first STEM Like a Girl workshop, I wasn't very interested in science at all. My mom started to invite my friends and took us more often, and then I began to like science. I enjoyed the experiments that I could do at home with household things. My aunt Maria took me to my last workshop, and we had fun. I was pretty sad when I was going to turn eleven because that meant I would no longer be able to attend the workshops. Now that I'm older, I did a ChickTech camp where volunteers shared their career paths in technology. I got to learn how to design and print a 3D treasure chest.

Why do you like doing STEM activities?
I like doing STEM activities because it involves everything around us. In my Ready, Set, Code classes at the library, the volunteer engineers set up the classes so we thought we were just playing, but really we were coding. It was a fun time.

What is your favorite STEM project you've ever done?
My favorite activity was when I got to see how our DNA was visible in a tube filled with liquid. At my aviation camp, I also really liked that I got to go inside different types of aircraft. We got to use computer flight simulators, which was fun because I could pretend I was a pilot.

When an experiment or design doesn't turn out how you expected, how does that make you feel?
When an experiment doesn't turn out how I expect it to, I feel frustrated, but I realize I can get through it and redo it. I just have to try again.

What other activities make you feel courageous, confident, and bold?
Aviation activities make me feel confident because I enjoy airplanes and flying. I remember touring the area where the mechanics were fixing their helicopters, and the tour guide asked if anyone wanted to be a mechanic. I

said, "I AM A MECHANIC!" This was funny to the group but very enlightening to my mom.

Who do you look up to?
One of the leaders at aviation camp, Becky, who built her own airplane by using twenty-two pages of drawings. It only took her seven years to build. She was kind enough to give me a ride in her two-passenger aircraft.

What are some ways girls can do STEM activities at home?
I like to take small electrical appliances apart to see how they work. The first appliance I took apart was our old toaster. After taking it apart I went online to see videos of people explaining the different components and how the toaster works. Ask your relatives or friends to give you their old electrical household items, and use the tools your parents might have in the garage, and see what you can learn!

What advice do you have for other girls?
To just try it because you will never know if you like it. I thought I didn't like science, aviation, and engineering, and now look!

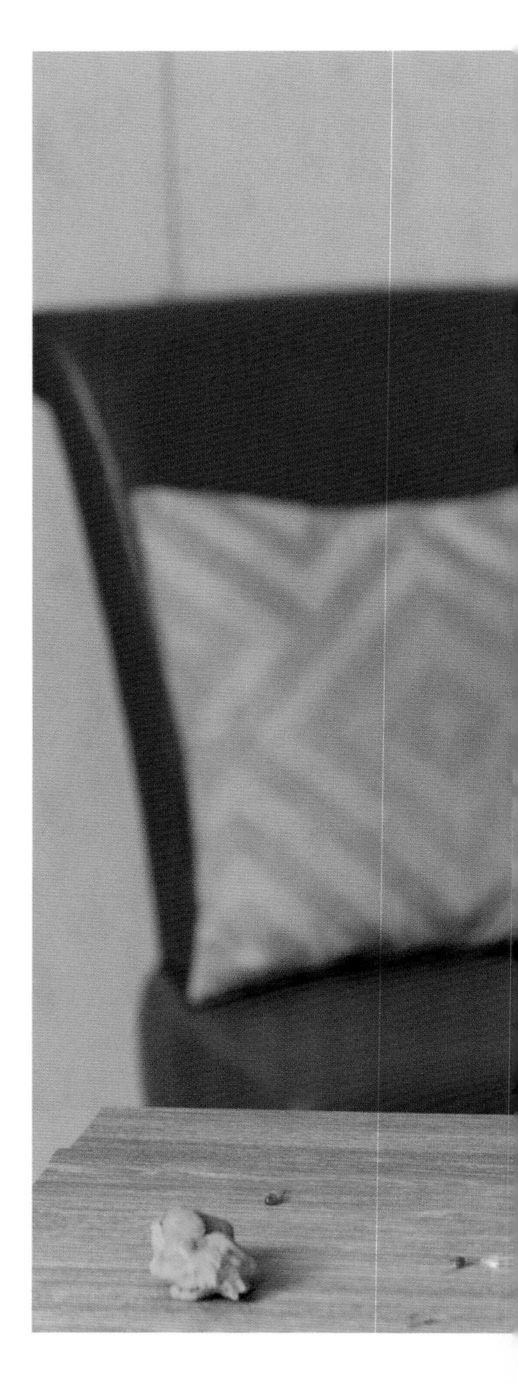

Krista, age 9

What are some things you want us to know about you?
I like drawing. My favorite thing to draw is anime characters with a stylus on my tablet. I love horror books, and I like to read. I also really like playing Minecraft, especially educational and survival modes. It's like technology because you have to coordinate how you get to a single block. You also do chemistry and math in educational mode and can experiment with it in lots of different ways.

Where do you do STEM?
I do STEM at the local science museum. I do it in my room, where I get to read books about math, science, technology, engineering. I also learn about STEM everywhere, like from a friend, my parents, a book, or even TV.

When did you first become interested in STEM?
I think it was the first time STEM Like a Girl came to the makers' fair at my school and did fun projects with us. I really enjoyed it. I was about six or seven years old then, maybe even five.

Why do you like doing STEM activities?
Because you get to experiment with things you haven't done before, and you can keep trying over and over again. Also, you get to do many different things, and you can get your friends to want to do STEM too. It's hard to do everything on your own sometimes, so it's better when you have friends to help you. Working with friends gives you way more ideas.

What is your favorite STEM project you've ever done?
One of my favorites was designing a city where I had to try using gardens and parks to make sure water didn't flood the houses and that pollution didn't get washed into a river. It was really fun to do.

When an experiment or design doesn't turn out how you expect, how does that make you feel?
I feel happy because I get to try it over and over again. Even when I do get it right, I like to see if there is an easier or different way to do it.

What other activities make you feel courageous, confident, and bold?
One of them is drawing. One of my favorite things to draw is anatomy, because you get to use a lot of lines and shapes to draw the face and body. Also playing Minecraft because it teaches you skills, and I'm good at it.

Who do you look up to?
My family members and my friends, especially my mom. She has her own business and works really hard at it.

What are some ways girls can do STEM activities at home?
One thing to do is to read a lot about STEM. You can also talk to your friends and think about all the different categories of STEM to learn about. And you can also make inventions in your everyday life.

What advice do you have for other girls?
To keep trying and trying until you get it right. If you want to, you can keep trying until you think it's perfect.

CRUMBLING CAVES

DIFFICULTY LEVEL

How many chocolate chips can you safely remove before the cookie crumbles?

MATERIALS

Several hard chocolate chip cookies
Several soft chocolate chip cookies
Tray or paper plate
Various digging tools (for example: plastic spoon, paper clip, toothpick, craft stick)

METHOD

1. Place one hard cookie and one soft cookie on your tray or paper plate. Label them "hard" and "soft" accordingly.

2. The goal of this experiment is to use your digging tools to remove the chocolate chips from the cookie without the cookie crumbling or falling apart. Examine the digging tools and determine which might work best for each cookie. Why would you choose one tool over another?

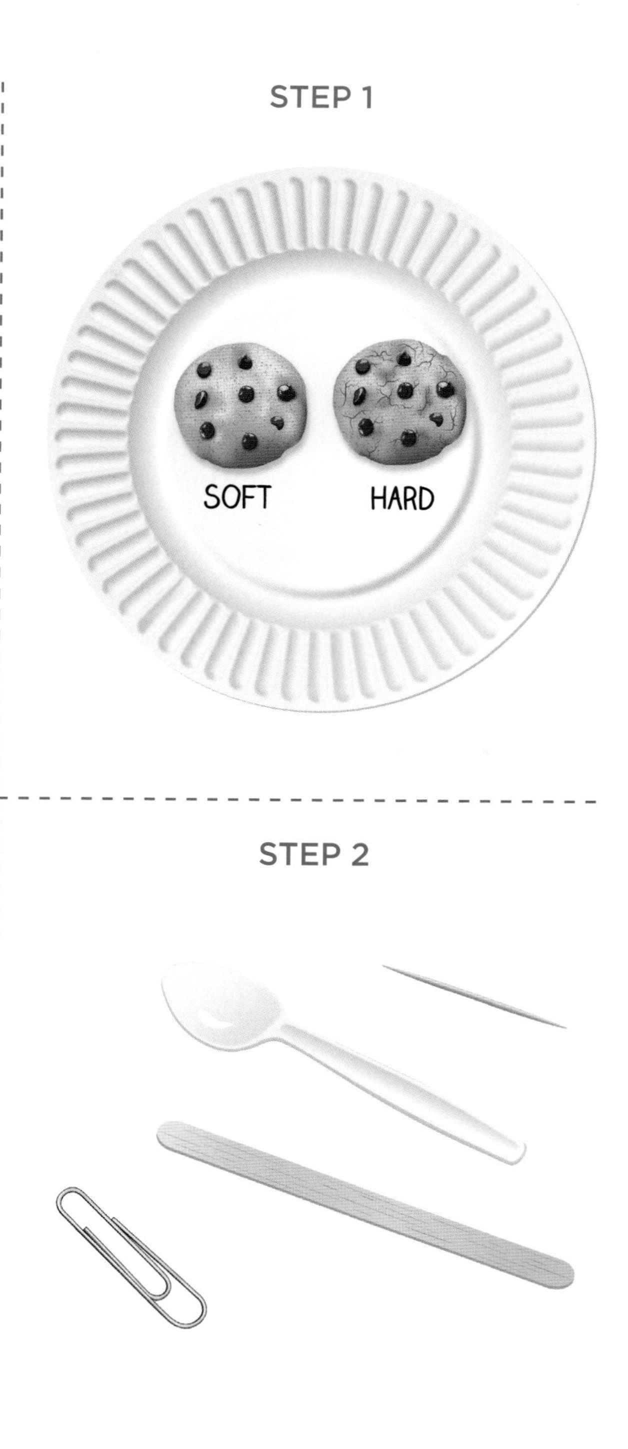

3. Look at each cookie and make a prediction about how many chocolate chips you think you will be able to remove without making the cookie crumble into pieces. Record your prediction on your experiment test sheet.

4. Select one tool to try first, and begin mining for the chocolate chips in each cookie. Make observations about how the tool works and how many chocolate chips you are able to safely extract. Record your results on your sheet.

5. When you are done mining each cookie, replace the used cookies with new ones, and select a different digging tool to try. Repeat the experiment, and make observations comparing each tool.

6. Which cookie type was easier to mine? What different tools or techniques were necessary for each cookie type? How do your results compare to your initial prediction?

STEP 4

STEM APPLICATION

Mining engineers, earth scientists, and environmental engineers (collectively called "miners") all work together to extract resources and artifacts from the Earth. Coal, oil, diamonds, iron, gold, and fossils are some examples of things miners dig for. Their goal is to help workers extract the largest amount of material from the ground while creating the least amount of damage to the environment. They also need to keep the workers safe, so they continuously monitor the ground around where they are digging to make sure it's stable. This can be quite a challenge!

Engineers and scientists also need to select the appropriate tools for the ground conditions they are digging in. They need to keep in mind the type of resource they are trying to extract. One tool might work well in sand while digging for fossils but would have a hard time extracting gold from inside a rock cave.

Lily, age 11

What are some things you want us to know about you?
I like to make wacky inventions, like last month I made a chair swing from a swing and a chair. The chair swing was really awesome because you could flip on it while sitting in it! I also found a way to make a seat belt out of the swing. To make the swing, I threaded the chain of the swing through the arms of the chair. Then I used the seat of the swing as a seat belt while swinging.

Where do you do STEM?
I do STEM in camps and a lot of it in school.

When did you first become interested in STEM?
In third grade we had a class called STEM, and we were solving problems. One problem was that our class had to create a prototype house that could withstand floods. I found it really fun because I love to design things like houses, interior spaces, structures, and prototype bridges, like we did in third grade. I love to design things.

Why do you like doing STEM activities?
STEM lets me stretch my mind and helps me design things. I like solving problems. Like at a workshop I got to design a tower that was high and could withstand an earthquake. I made it out of straws and marshmallows, and I remember I failed so much, but in the end I had the tallest structure.

What is your favorite STEM project you've ever done?
I liked it when my friend and I built a wind sculpture with moving parts at a workshop. I enjoyed this project because of all the fun and creative designs and wind sculptures I saw others building. I just loved solving the problem of it not tipping over. It tipped over so many times, but in the end we learned that we just needed to extend the base.

When an experiment or design doesn't turn out how you expected, how does that make you feel?
I feel disappointed, but I also feel good because I learned something and I can improve what I failed at.

What other activities make you feel courageous, confident, and bold?
When I am playing sports it makes me feel strong, or while playing dodgeball and I get someone out.

Who do you look up to?
I look up to Coco Gauff because she pursued her dreams and became a pro tennis player at the age of fourteen.

What are some ways girls can do STEM activities at home?
You can find a science project to do, invent something, and learn something new.

What advice do you have for other girls?
If you fail, it's good. It means that you're learning!

Molly, age 10

What are some things you want us to know about you?
I love to read books and play on my tablet. My favorite nontablet activity is to build Lego things. I like the kits, but I really like making my own builds.

Where do you do STEM?
I do STEM at school, at home, at STEM Like a Girl classes with my mom, and my parents take me to any science museum or aquarium they can find when we travel.

When did you first become interested in STEM?
I remember doing a crystal-growing kit at home, and it was interesting to me that using some of the same ingredients in a different way created different crystal formations and different colors. Space science is my favorite right now. I read and watch everything I can about NASA and space.

Why do you like doing STEM activities?
They are just fun. With my Legos, I like trying to think of a creation and see if I can make it. Most times I can get close, but it usually turns out better than I thought it would. When I look at the crystals I grow or look at things in my microscope, I like to see what built those. What chemicals or natural things came together to create what I am seeing.

What is your favorite STEM project you've ever done?
My favorite STEM project that I have done was using coding robots, like Sphero, at school. We set up an obstacle course and then had to write code so that the robot could move through it without bumping into any of the objects.

When an experiment or design doesn't turn out how you expected, how does that make you feel?
Sometimes it makes me happy because I can do it all over again and see if I did it wrong or if my hypothesis was wrong. Sometimes it is disappointing though.

What other activities make you feel courageous, confident, and bold?
I feel courageous when I try new things and I see that they aren't as hard as I thought, like when I went paddleboarding. It was so much easier than I thought. This made me see why it is important to try new things. I am courageous and confident when I dance and go onstage to perform. I have practiced the dances over and over, and I feel good about how well I know them before I go onstage to share them.

Who do you look up to?
I look up to my mom because she is smart. She is a teacher and had to go to college and get degrees in order to do that. And Ada Lovelace because she said, "Your best and wisest refuge from all troubles is in your science." This reminds me that science can make you feel safe.

What are some ways girls can do STEM activities at home?
Girls can do the same activities that boys can! They can get kits if they just want to follow directions and see what happens, or they can try to answer their own questions by setting up experiments and trials. I even have a book that gives me ideas of experiments I can try, so they could get one of those.

What advice do you have for other girls?
Don't be nervous or scared about things. You have to try new things in order to decide if you like them or not. If you don't try anything, you can't have an opinion about it.

SQUISHY SOAP

DIFFICULTY LEVEL

Harness the power of hydrogen bonds to make your own cross-linked jelly soap.

MATERIALS

Safety glasses
Gelatin powder (unflavored)
Water
Liquid body soap
Food coloring (optional)
Measuring spoons and cups
Small bowl
Spoon
Silicone molds, metal muffin tray, or small plastic cups
Craft stick

METHOD

1. Put on your safety glasses. When working with chemicals, it is always important to protect your eyes!

2. Place 2 tablespoons of gelatin powder (two 0.25 oz. packets) into the bowl.

3. Carefully boil approximately 1 cup of water.

4. Slowly add 1/3 cup of the boiling water to the bowl containing the gelatin, and carefully stir with a spoon until the gelatin is dissolved. Use the back of the spoon to break apart any large pieces of gelatin.

STEP 4

5. Add 1/3 cup of liquid soap and 1–2 drops of food coloring (if using) to the bowl, and gently stir with a spoon to combine. Try to avoid getting too many bubbles.

6. Carefully pour the mixture into the silicone molds or cups.

7. Check your mixture after an hour by lightly pressing it with a craft stick. What do you observe? Record your observations on your experiment test sheet. Over time, the cross-linking process transforms your squishy soap from a liquid to a solid.

8. After several hours, the cross-linking process should be complete. Use a craft stick to loosen the form from the mold or cup so it's easy to remove. Wrap the squishy soaps individually in plastic wrap until you use them. Note: If you live in a warm environment and don't plan to use the soap right away, store them in the fridge.

STEP 6

STEP 7

9. When it's time for a shower or bath, unwrap one of your squishy soaps and use it like any regular bar of soap. Observe what happens as the squishy soap warms up in the hot water.

10. What happens if you change the amount of gelatin in your soap? Experiment with adding more or less gelatin to the water. Make observations about how the concentration of gelatin affects the feel and behavior of your squishy soap.

STEP 8

STEM APPLICATION

Gelatin comes from collagen, which is a protein made up of long strands of chemicals called amino acids. (Amino acids are the building blocks of proteins.) When dissolved in water, these gelatin strands like to interact with each other and stick together to form a three-dimensional matrix. This process is called *cross-linking* and the resulting gel is called a *hydrogel* (a cross-linked gel containing water). At room temperature, these interactions are stronger so the gelatin is more solid. However, when you heat the gel (like in the shower), the interactions are weaker, causing the gel to dissolve into a liquid.

There's another interaction happening between the water and the gelatin strands. Gelatin has hydrogen atoms as part of its chemical structure. Water is made of hydrogen and oxygen atoms. The hydrogen atoms in the gelatin interact with the oxygen atoms in the water. This forms weak bonds called *hydrogen bonds*. These bonds aren't true chemical bonds, but they are enough to increase the strength of the hydrogel. When you heat up the gel, these bonds become weaker and break apart.

Sophia, age 12

What are some things you want us to know about you?
I can play piano and have been playing since I was five years old. I really like playing music that is fast and loud! I also really like to play Minecraft and Summoners War during my free time.

Where do you do STEM?
I do STEM at home with some basic experiments, at school in science class, and at STEM Like a Girl.

When did you first become interested in STEM?
At first my mom signed me up for STEM Like a Girl, and I really liked all the activities there. One activity that was really cool was where you built a building and tried to make it withstand a mini earthquake. I kept experimenting and learned to make the bottom stable but keep the top small so it wouldn't fall over. I put it on a table that would shake my building like an earthquake would, and it stayed standing!

Why do you like doing STEM activities?
I like doing STEM activities because I like building things and learning how things work. It's really fun and exciting to create new things.

What is your favorite STEM project you've ever done?
My favorite STEM project I've ever done is a project where I made a small car out of some simple materials. I used things like straws, cups, and tape, and I had to try to see which materials would make it move the fastest.

When an experiment or design doesn't turn out how you expected, how does that make you feel?
When an experiment or design doesn't turn out how I expected, it makes me feel more determined, and I try again until I make it right.

What other activities make you feel courageous, confident, and bold?
Playing piano makes me feel courageous, confident, and bold. I like to go to festivals and play for other people. I was nervous playing in front of so many people, but once I did it I felt strong.

Who do you look up to?
I look up to my mom because she teaches me a lot of things.

What are some ways girls can do STEM activities at home?
A way to do STEM activities at home is to look up some simple activities that use materials that you have at home. To make it more fun, you can do them with siblings, parents, or friends.

What advice do you have for other girls?
Try your best, and even if it doesn't turn out how you expect, keep trying. Each time, you will learn different concepts and find out a new way to get it done.

Maya, age 10

What are some things you want us to know about you?
I like to play soccer and volleyball, and I love to paint. I really like to paint landscapes using acrylic paints.

Where do you do STEM?
I do STEM at school, home, Girl Scouts, summer camp, and the science museum.

When did you first become interested in STEM?
I got interested in STEM when my librarian suggested STEM Like a Girl, and I did my first workshop. It was really fun because one of my friends did it with me, and we made another friend there. We did a DNA project and it was really cool.

Why do you like doing STEM activities?
I like doing STEM because I like to build things and mix different things and make something cool.

What is your favorite STEM project you've ever done?
My favorite STEM project that I have done is building a bridge in third grade out of cardboard, Popsicle sticks, and string. Currently, for camp, I am building a hotel that you can take apart. We are using cardboard, Popsicle sticks, and hot glue.

When an experiment or design doesn't turn out how you expected, how does that make you feel?
It makes me feel frustrated. When that happens, I try calming down, and then I try again. When it finally works, I feel happy.

What other activities make you feel courageous, confident, and bold?
I feel courageous and confident when I am making art and playing soccer.

Who do you look up to?
I look up to my fourth-grade teacher because she taught me a lot and made schoolwork fun.

What are some ways girls can do STEM activities at home?
You can build stuff out of recycled materials. Making slime and oobleck is really fun. You can also make a catapult out of Popsicle sticks, hot glue, a rubber band, and a bottle cap.

What advice do you have for other girls?
Be kind to others, try not to give up, and have fun.

Siuma, age 12

What are some things you want us to know about you?
I go to middle school and love swimming, biking, basketball, and hanging out with my friends.

Where do you do STEM?
I do STEM in school, at after-school programs, and at home.

When did you first become interested in STEM?
I became interested when I took my first STEM Like a Girl workshop. I liked it so much that I decided to continue doing it.

Why do you like doing STEM activities?
When I do STEM activities, it makes me curious about how things work. In every experiment that I do, I come across something that doesn't quite pop. My favorite thing is when I figure it out and say, "Oh…I got it!"

What is your favorite STEM project you've ever done?
I tried to package up a real egg so that it wouldn't crack on impact when dropped from a ladder. I used packing peanuts, felt, a cardboard box, string, and about fifteen balloons. I thought the balloons would cushion it more, but overall it turned out really well and the egg didn't crack.

When an experiment or design doesn't turn out how you expected, how does that make you feel?
It makes me feel perplexed, because when it doesn't turn out right, I wonder what went wrong. When that happens, I would try to examine what I first did and then see what I can change.

What other activities make you feel courageous, confident, and bold?
An activity that makes me feel courageous and bold is swimming because I have been swimming competitively for several years and that makes me feel more confident.

Who do you look up to?
I look up to the Nobel Prize–winning scientist Linda Buck because she discovered how the olfactory system works, which is your sense of smell. Like her, I want to discover something essential to life.

What are some ways girls can do STEM activities at home?
Be curious and look up fun STEM activities online to do while at home.

What advice do you have for other girls?
If you ask questions and try your hardest, you will always succeed.

PUDDLING PENNIES

DIFFICULTY LEVEL

How many drops of water can you fit on a penny?

MATERIALS

Safety glasses
Water
Liquid dish soap
Isopropyl alcohol (rubbing alcohol)
3 pennies
3 small cups
Pen or pencil
Measuring cup
Measuring spoon
Rimmed baking pan
Pipette or dropper

METHOD

1. Put on your safety glasses. When working with chemicals, it is always important to protect your eyes!

2. Prepare your solutions:

Label your first cup "water," and fill it with 1/4 cup water.

Label your second cup "soap water," and fill it with 1/4 cup water and 1 tablespoon of liquid dish soap. Gently stir with a spoon.

Label your third cup "isopropyl alcohol," and fill it with 1/4 cup isopropyl alcohol.

STEP 2

3. Place three dry pennies in the baking pan. Make sure they all have the same side facing up as this will affect the results.

4. Using a pipette or dropper, add one drop of water to the center of the first penny's surface and observe what happens.

5. Continue to add water to the same penny, one drop at a time, counting the number of drops as you go.

6. What observations can you make as you keep adding drops? Does the water flow out along the surface of the penny, or does it stay puddled together?

7. See how many drops you can add until the water falls over the edge of the penny. Record this number and your observations on your experiment test sheet.

8. Use the next two pennies to repeat the experiment with soap water and isopropyl alcohol. How many drops can you add each time?

STEP 4

STEP 6

STEM APPLICATION

The phenomenon at work here is called *surface tension*. Water is made of molecules that attract one another quite strongly, forming *hydrogen bonds* (hydrogen atoms binding with oxygen atoms). Because of this, the molecules on the surface of the water cling together, or puddle, creating tension. It requires a lot of force to break these hydrogen bonds. As you add drops, instead of spilling over the edge, the water forms a dome until there is enough force from gravity to overcome the surface tension.

When soap is added to the water, some of the water molecules bind, or attach, to the soap molecules instead of to each other. This disrupts the hydrogen bonds and reduces the strength of the water molecules' attraction to one another. Because of this, the surface tension in soapy water is much weaker than in pure water. The soap water spills over the penny's edge sooner than it does with pure water because the force of gravity overcomes the surface tension faster. The more liquid soap you add to the solution, the lower the surface tension, and the fewer drops stay on the penny. You can test this hypothesis by adding another squirt of soap to the soap water solution and repeating the test.

Isopropyl alcohol also forms hydrogen bonds, but they are weaker than the hydrogen bonds in pure water. The surface tension of isopropyl alcohol is less than half the surface tension of pure water. So, as with the soap solution, fewer drops of the alcohol solution stay on the penny before gravity takes over and the liquid spills off.

Isabella, age 11

What are some things you want us to know about you?
I live with my parents, two older sisters, and my dog named Buzz. I like doing all types of things. My favorite thing to do is sing. Other hobbies I enjoy are sewing, robotics, and baking.

Where do you do STEM?
At home, at school, at camps, and at the public library. I have done a camp in the past called Camp Invention. At school I did many after-school programs, and at the library I did a five-day camp called STEM You Can!

When did you first become interested in STEM?
When I was little, my mom bought my two older sisters a big Lego set for Christmas. While my sisters would build, I would play with the extra pieces they weren't using and would make something different. Ever since, I like fixing and tinkering with broken things like messed-up clothes and old toys.

Why do you like doing STEM activities?
I like doing STEM activities because they are really fun, and the things that you learn from STEM can be used in everyday life. For example, some of the hobbies I enjoy are sewing and baking, which require STEM. Baking requires math, and sewing requires both math and engineering.

What is your favorite STEM project you've ever done?
It would have to be fixing up my cousin's old skateboard and making it my own. I was looking for a skateboard in stores, but I didn't like any of them. I went to my grandparents' house one day and I found my cousin's old skateboard. The grip tape was peeling, the trucks were too loose, and it did not fit my style. However, it had potential. So, with my dad's help, I sanded and painted it, put new grip tape on it, and put on new trucks, bolts, and wheels. Finally, I had a skateboard that was a perfect fit. I was proud of myself for making something I could actually use.

When an experiment or design doesn't turn out how you expected, how does that make you feel?
It makes me a bit disappointed. But giving up won't do anything. For example, when I sew, I sometimes ruin a stitch or make the wrong measurements. We all make mistakes, but we can always fix them.

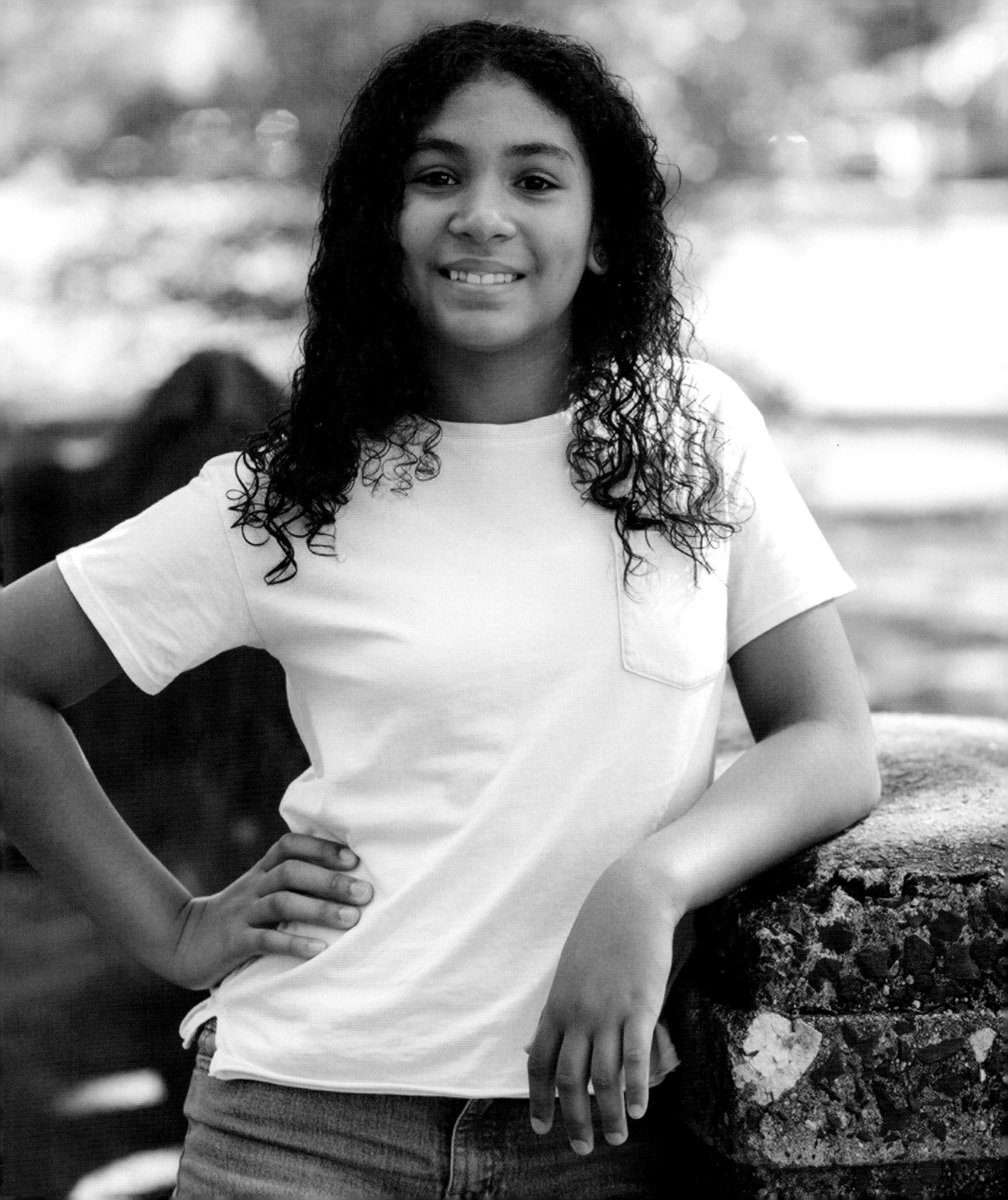

What other activities make you feel courageous, confident, and bold?
I enjoy making desserts, creating, singing, sewing, and acting.

Who do you look up to?
I really don't have a person I look up to. However, if I see something old, it inspires me to make it new, modern, and nicer.

What are some ways girls can do STEM activities at home?
Remember, not all STEM is science experiments and over-the-top investigations. I encourage girls to try solving problems in their environment. For example, if there are tangled wires hanging around, try inventing something to organize the wires.

What advice do you have for other girls?
My advice is that no idea is a bad idea. Why? Because that so-called bad idea is one day going to have an impact on someone else's life for the better.

Audrey, age 9

What are some things you want us to know about you?
I'm very tall for my age. I'm a fast runner and I really like running. I did a running camp this summer. I really like climbing trees, and once I climbed seventy feet with my friend Emma. I like playing defense in soccer. I really like ducks, especially mallard ducks. They are my favorite animal.

Where do you do STEM?
I do STEM at my house with my mom and outside in my backyard.

When did you first become interested in STEM?
When I was three, I started learning about outer space. Ever since then, I've learned a lot about being an astronaut, the moon, and planets. My sister and I want to be the first sisters on the moon. I also like getting STEM kits, which have fun projects that come with instructions in a box. My favorite one to build was my own secret-agent kit with a periscope to see backward.

Why do you like doing STEM activities?
I like doing STEM activities because they are fun and they keep your mind working. Sometimes you have to try a project again before it's how you want it, so they keep me busy.

What is your favorite STEM project you've ever done?
I'm not finished with it yet, but right now I'm building a playhouse with my dad that we can sleep in. It's six feet by eight feet. We are going to put the joists in today. We are going to put panels on the ceiling so that I can make curtains and make different-shaped rooms, like a triangular room.

When an experiment or design doesn't turn out how you expected, how does that make you feel?
I sort of feel sad, but then I learn from my mistakes and get to try again. Scientists never succeed on the first try, and even if they do succeed they keep making changes and try again and again to learn more.

What other activities make you feel courageous, confident, and bold?
When I got third place in the spelling bee at school. When I did my first cartwheel. When my grandma told me that I would be a good ranch hand because I like helping with the hay on their ranch and she thought I was a hard worker.

Who do you look up to?
My aunt Sonya because she's superstrong, and I want to be strong too. And Sally Ride, because I want to be the first person to land on a planet, and she was the first American woman to fly on a space mission.

What are some ways girls can do STEM activities at home?
Do science fair at school, build a bridge, or build a mini model of a house. You can also look up projects on the internet and try them—like chemical reactions, making slime, making milk plastic, making your own compass out of a cork and a needle and water in a container, making a tornado in a bottle, or comparing natural and synthetic dyes.

What advice do you have for other girls?
Try not to get frustrated with yourself. If you get frustrated, think about what went wrong and try something different. Take a break and drink some water. If life gives you lemons, make lemonade.

SOUND SCIENCE DIFFICULTY LEVEL

Make a harmonica with a couple of craft sticks, rubber bands, and a straw.

MATERIALS

Safety glasses
2 jumbo craft sticks
1 thick rubber band
(size #82 works well)
2 small rubber bands
(size #16 works well)
1 straw
Scissors

METHOD

1. Put on your safety glasses. When working with chemicals, it is always important to protect your eyes!

2. Wrap the thick rubber band lengthwise around one craft stick. You'll really have to stretch it, so use your muscles!

3. Cut two pieces of straw, each approximately 1 in. long.

4. Place one piece of straw *between* the large rubber band and the craft stick, and rest the other piece *on top* of the rubber band.

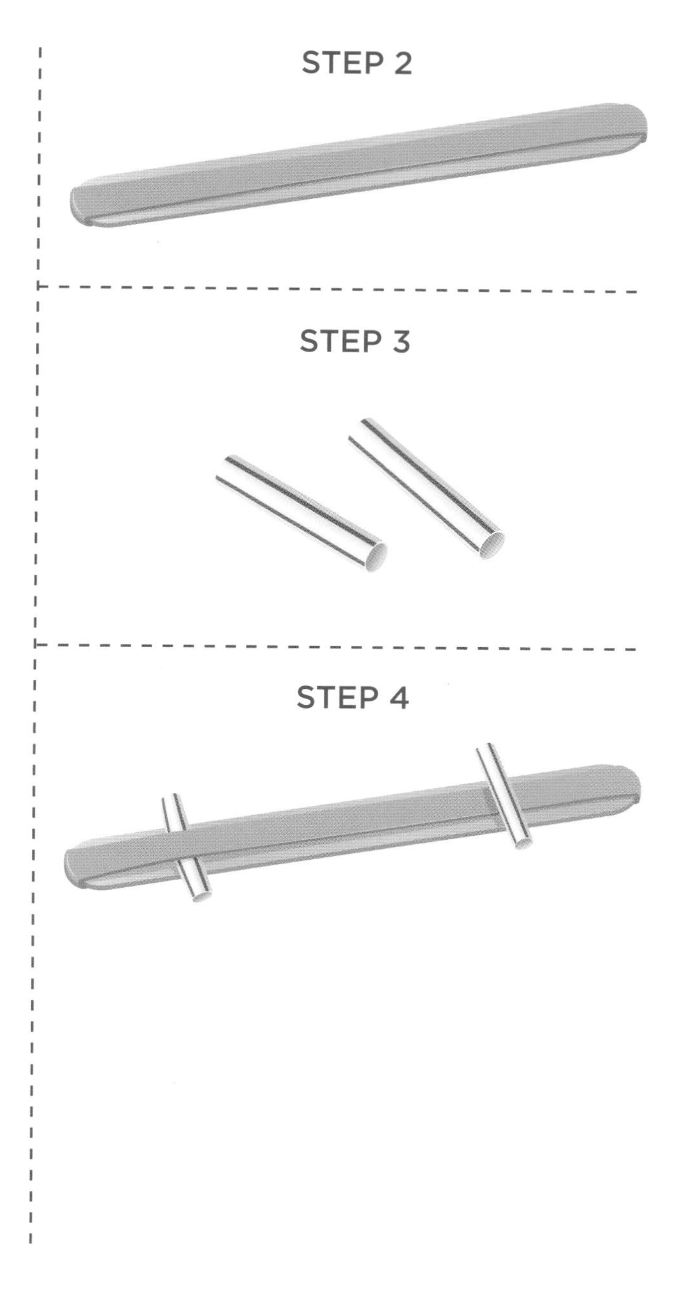

5. Sandwich it all together with the second craft stick.

6. Wrap each end tightly with the small rubber bands. The straw pieces can be located wherever you want them along the length of the craft sticks, but they should be placed between the small rubber bands.

7. Put your mouth on one long side of the craft sticks and blow into your harmonica. You aren't blowing into the straws but into the space between the straws.

8. Make observations and record them on your experiment test sheet. Can you play music with your harmonica? How does the sound change if you move the straws closer together or farther apart?

STEP 6

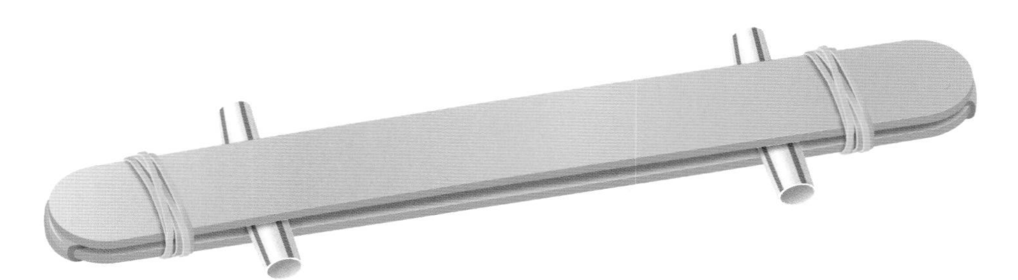

STEM APPLICATION

When you blow on the harmonica, air passes through the space between the two craft sticks. Since the straw pieces separate the thick rubber band from the craft sticks, the air causes the thick rubber band to move up and down between the sticks. This movement is called *vibration*. The sounds that you hear are the result of the *energy* that is produced when the thick rubber band vibrates.

In order to change the sound produced by the harmonica, you need to change the *frequency of oscillation* of the thick rubber band. Frequency of oscillation simply means the number of oscillations (or vibrations) in a set period of time. When the straws are far apart, the rubber band takes longer to move through one oscillation. This results in a lower-pitched sound. When the straws are close together, the rubber band oscillates at a much faster frequency, resulting in a higher-pitched sound.

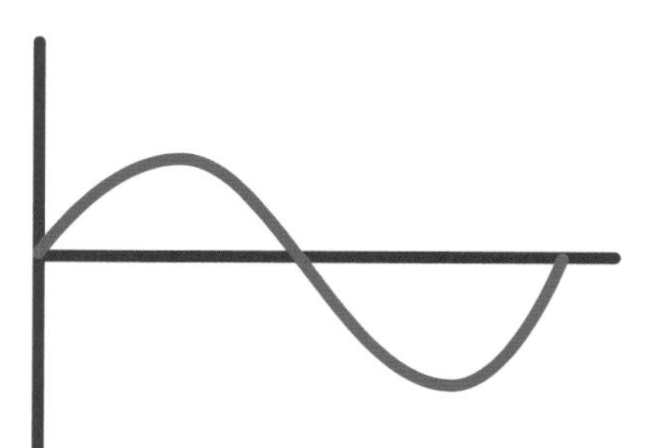

Lower Pitch

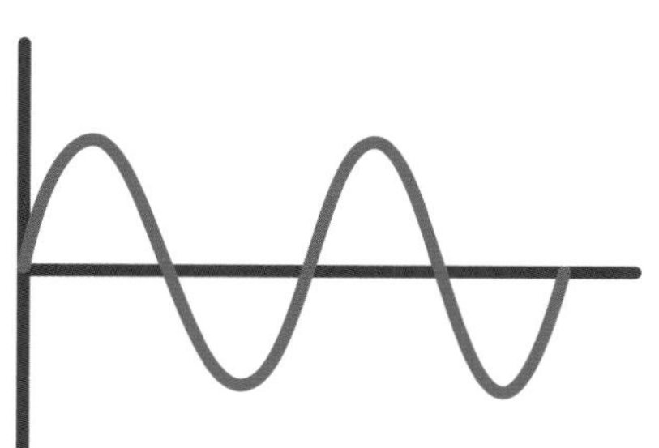

Higher Pitch

Maisey, age 11

What are some things you want us to know about you?
I want to be a marine biologist when I get older. I love doing STEM projects, especially designing my own inventions to solve problems I see. My favorite subjects are math and science. I also have learned all about the patent process while applying for a patent for my own invention!

Where do you do STEM?
I do STEM activities at my school in STEM class, at an after-school STEM club, at Girl Scouts, at summer camp, and at home.

When did you first become interested in STEM?
I first became interested in STEM at school. For icebreakers, we would do small STEM challenges. To me, these felt like puzzles, and I soon fell in love with them.

Why do you like doing STEM activities?
I like doing STEM activities because I love challenges and I love learning in a hands-on format.

What is your favorite STEM project you've ever done?
My favorite STEM project I've ever done would probably be when I designed the invention that I've now applied to have patented. I love it because I am passionate about the reason I made it: to help save a marine species called the North Atlantic right whale. One STEM project that I loved doing was when I made a radio in the backyard. I used paper, foil, a paper-towel tube, my bike, paper clips, and a tool that helped control the electric flow. Once I set it up correctly, I could listen through my headphones and try to find a nearby radio station I liked.

When an experiment or design doesn't turn out how you expected, how does that make you feel?
I often feel frustrated, but mostly I become curious and want to understand the problem so that I can try to create a solution.

What other activities make you feel courageous, confident, and bold?
I love reading, rock climbing, skiing, coding, playing with my dog, camping, and crafting.

Who do you look up to?
I look up to important women in marine biology, including Sylvia Earle, Maria Mitchell, and Cindy Lee Van Dover.

What are some ways girls can do STEM activities at home?
Baking, creating, and coding are all forms of STEM that are easy to do at home. Also, taking apart old broken machines can provide awesome parts for your next STEM project!

What advice do you have for other girls?
Find a simple problem around your house, and come up with a machine to solve that problem. It can be as simple or as complicated as you'd like. Be sure to use household materials. You'll be surprised what you can do with some simple materials!

Luchia, age 11

What are some things you want us to know about you?
I am a volleyball player and I love the sport. I started playing when I was nine years old on my school team and then decided to try out for a more competitive team. Even though I was young, they saw something in me and gave me the opportunity to join their team. I went to three days of their tryouts, was the first one on the line every day, and never gave up no matter how tall everyone else was. I was chosen to be a part of the team and I became their libero [a special defensive player]. I have also been a Girl Scout for six years. I have an older and a younger brother. I also love making slime! My dad helped me recently start my own YouTube channel, where I use my skill with STEM to show kids how to make crafts and express their feelings using STEM by making slime, creating homemade sketchbooks, and so much more that requires experimenting and measuring everything you use.

Where do you do STEM?
I started STEM when I was three years old. My parents would take me and my brothers to Home Depot activities on Saturday morning, where we would build birdhouses or pencil boxes. I also do it in Girl Scouts, at school, in the science fair, and at the library.

When did you first become interested in STEM?
I have always done STEM. I just never really knew what it was until my mom started putting me into the actual activities and I started doing STEM-based projects at school.

Why do you like doing STEM activities?
I knew I really loved STEM when I started making slime. I love experimenting with the amounts of saline solution and baking soda (activator), color, gel, clay, and beads to make it crunchy and whatever else it needs to make sure that it becomes the texture I want. It is also all about measurements and mixtures. That is when I realized that STEM was fascinating to me and that girls can rock this STEM thing!

What is your favorite STEM project you've ever done?
I love the slime experiments I do, and they help calm me. When I feel anxious or feel my stomach in knots or am annoyed with my brothers or angry at something or even sad, my slime and crafts keep me calm. I know what texture or mixture will make me feel better. I know that I want that certain click, clack, or crunch of the slime to make my mind calm down. This is when STEM comes to the front lines. I need to make sure all my measurements are okay. If not, the slime will stiffen or not be the perfect texture I want.

When an experiment or design does not turn out how you expected, how does that make you feel?
I will just do it again. It is so much fun that doing it again really doesn't bother me.

What other activities make you feel courageous, confident, and bold?
Volleyball makes me feel very confident. I am the shortest and youngest, so when people look at me they think I can't do it or I can't score or serve a point. Once people see me on the court, they see that I can get low to hit the balls that others spike. I have been able to get us some winning points.

Who do you look up to?
I look up to my grandmother, because whenever she gets hurt or is sick, she just keeps on going. Nothing stops her! I want to be just like that.

What are some ways girls can do STEM activities at home?
My dad says STEM is anything that gets your mind thinking, gets you involved in measurements, and gets you experimenting and building. It is about trying to get you to become more of a critical thinker, which means to really be able to analyze, experiment, talk about things. So really anything you do to get your brain working can lead you to STEM.

What advice do you have for other girls?
My mom is the leader of my Girl Scout troop, and she has forever taught my troop one thing: "DO IT LIKE A GIRL." There is nothing a boy can do that a girl cannot. There is nothing that I cannot do. I can accomplish anything I put my mind to. Never give up. As frustrating as it is, as scary as it seems, do not stop trying. The only time that you will not accomplish something is when you quit or do not even try!

Cece, age 9

What are some things you want us to know about you?
I love playing soccer with my best friend, Emma. I also love going swimming with my cousins. I have a sister named Lena, and sometimes we play together.

Where do you do STEM?
I do STEM in school. I have a class in school called iSTEAM where we build things and do experiments. For example, one time we had to cut a piece of paper into as many rings as we could that were all connected together. I had to try to make the rings skinny so I would have more paper. I also watch videos at home and try DIY experiments.

When did you first become interested in STEM?
I became interested in STEM when I had my first STEM class. I started learning about what we do in STEM and what it stands for. I loved doing all the experiments.

Why do you like doing STEM activities?
I like doing STEM activities because I like doing science and math. I am good at it, and my teacher makes it fun.

What is your favorite STEM project you've ever done?
My favorite STEM project that I have ever done is when, in my class, we had to make a miniature bed with only a certain amount of materials.

When an experiment or design doesn't turn out how you expected, how does that make you feel?
When an experiment or design does not turn out how I expected, I feel frustrated at first, but then it makes me determined to do it again until I do it right.

What other activities make you feel courageous, confident, and bold?
Other activities that make me feel courageous, confident, and bold are soccer and piano. I play them both. I feel good when I am out on the soccer field and running and scoring goals. I also love learning new songs on the piano.

Who do you look up to?
I look up to Ruth Bader Ginsburg because she was a strong woman! Even when she went through health issues, she never gave up. She was really important to our country.

What are some ways girls can do STEM activities at home?
Some ways girls can do STEM activities at home is to find some supplies at home and turn those supplies into art or sculptures. I like to look at YouTube videos for ideas.

What advice do you have for other girls?
I would tell girls you should never give up on anything! If something turns out how you did not want it to, don't give up, and just keep trying! The mistakes that you make can help you get better at it!

BLAST OFF DIFFICULTY LEVEL

How far can you shoot a paper rocket with a straw?

MATERIALS

1 flexible straw
Thin paper, like printer paper or origami paper
Thick paper, like construction paper or card stock
Ruler
Pencil
Scissors
Tape

METHOD

1. First, make the body of the rocket. To do this, cut your thin paper into a rectangle approximately 4 x 3 in. Roll the paper into a tube with an opening large enough to fit over the straw. You can decide which direction to roll it in. Slip the rolled paper over your straw to make sure it fits. Tape the paper so it doesn't unroll.

2. With the rocket body on the straw, blow on the end of the straw and observe what happens. Because the end of the rocket is open, the air escapes out of the top so the rocket has no power to blast off.

STEP 1

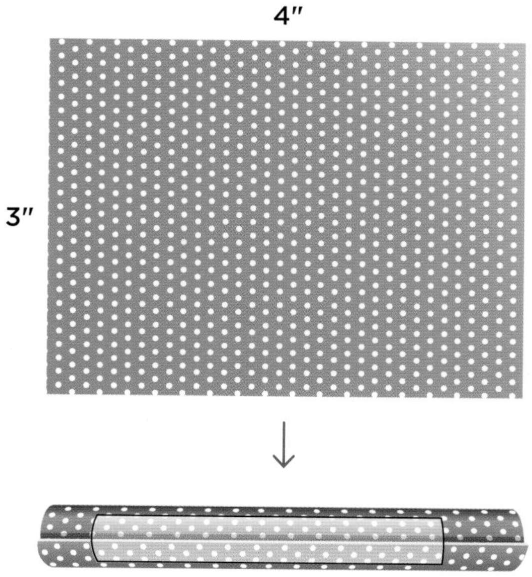

4″

3″

STEP 2

3. In order to make the rocket fly, you'll need to seal off the top of your paper tube. There are two ways you can choose to do this:

Simply tape one end of the paper tube shut

OR

Make a nose cone

To make a nose cone, cut a piece of the thicker paper into a triangle with sides measuring approximately 4 x 3 x 3 in. Twist it into a cone shape, and tape it together. You can trim the edges if needed. Tape the cone onto the top of the rocket body to seal it off.

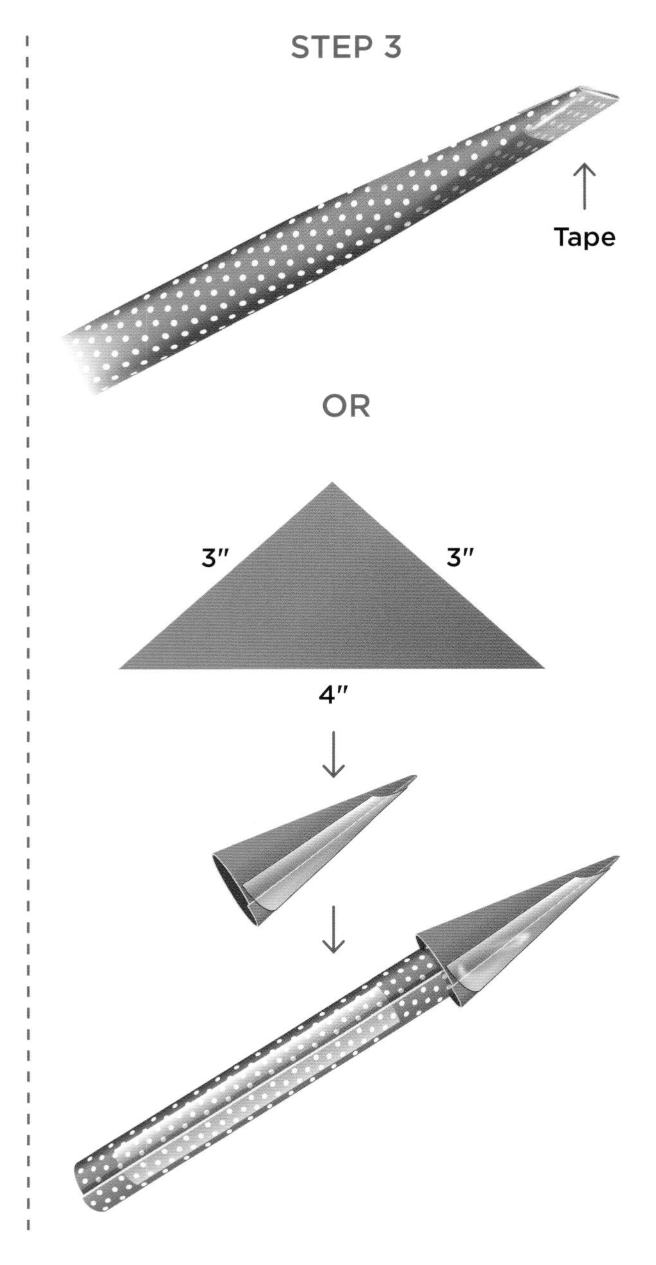

4. Think about a rocket ship or airplane. What else could you add to your rocket to help it fly? For example, you could include fins at the end of the rocket or wings along the sides. Use additional pieces of the thick paper and tape to add other design elements.

STEP 4

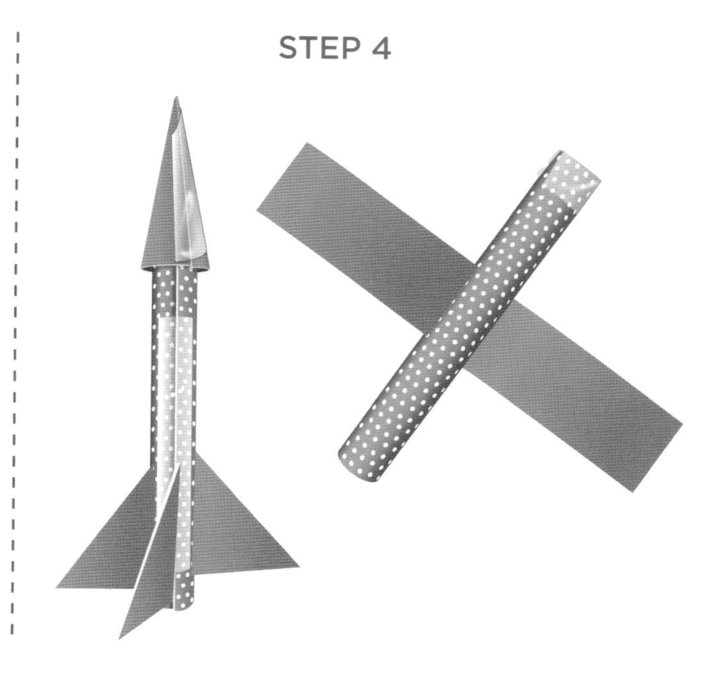

5. Now it's time to test the flight of your rocket. Place your rocket on the end of the straw and blow into the other end of the straw.

6. How far did your rocket fly? Did it fly straight? What modifications could you make that may allow better flight? Record your observations on your experiment test sheet.

7. Try testing out different designs, and challenge a friend or family member to see whose rocket can fly the farthest.

STEP 5

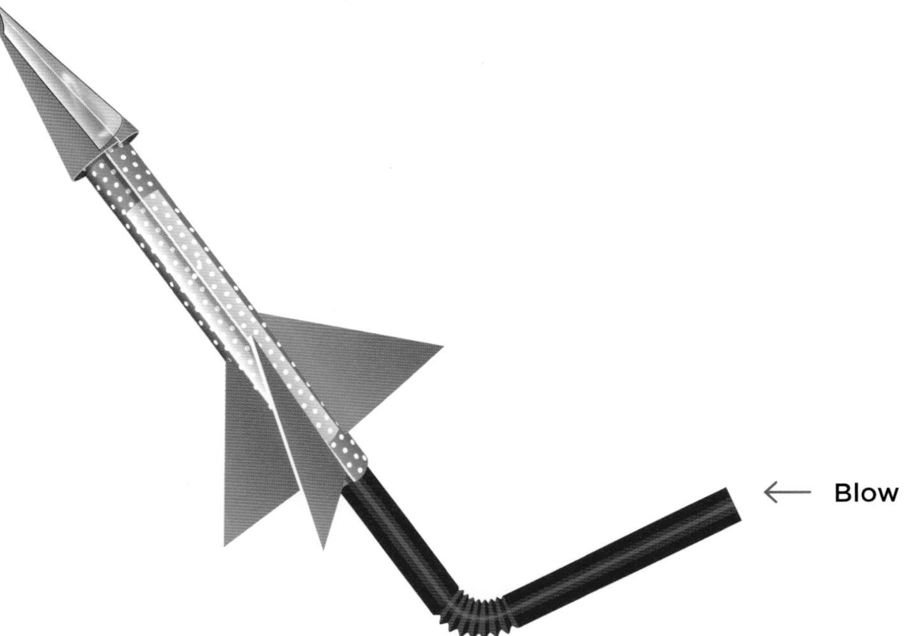

STEM APPLICATION

Aerodynamics is the study of how things fly and how air moves around a solid object. Everything that flies—including airplanes, rocket ships, and birds—are affected by the forces of aerodynamics. The more aerodynamic a flying object is, the better it will fly.

When designing an object that flies, there are several things engineers need to consider. *Drag* and *thrust* are two forces that work against each other. Drag is the force that slows down an object as it moves through the air. Thrust is the force that pushes something forward and keeps it moving. In order to keep an object in flight, the thrust force needs to be greater than the drag force. The *weight* of an object is also important for flight. If the flying object is within our atmosphere, the forces of gravity will be acting on it, pushing it toward the ground. A lighter object will remain in flight longer. *Stability* while in flight is necessary to keep the object moving in a straight direction or allow it to turn when needed. Certain design elements, such as fins or wings, can add stability to a flying object.

Ella, age 11

What are some things you want us to know about you?
I like playing sports, my favorite color is green, and I love reading and writing my own stories. I love to read and write mystery stories and realistic fiction.

Where do you do STEM?
I've gone to a STEM Like a Girl workshop. I also do projects at school and many experiments at home.

When did you first become interested in STEM?
When my mom first started introducing me to what STEM means and the different projects that go along with it. We have a makerspace at school with a bunch of random things, and we can go to that space and make anything. We can program robots and do so many cool things. I love building things with the STEM teacher at our school.

Why do you like doing STEM activities?
I like doing STEM activities because they let me do what I want to do and experiment with different materials.

What is your favorite STEM project you've ever done?
My favorite STEM project I've ever done was creating a chain-reaction project with a friend. Our goal was to get a ball into a cup without it touching the ground. We needed ramps, tubes, dominos—anything else we found around the house that could help us. It took many tries, but we finally got the ball into the cup.

When an experiment or design doesn't turn out how you expected, how does that make you feel?
I feel frustrated at first, but then I remember that making mistakes is just a part of life. You usually learn from mistakes so you can make improvements so it works better next time.

What other activities make you feel courageous, confident, and bold?
I feel confident by being on the basketball court, writing down my ideas to share with the world, creating a song even if I keep it private, and just being myself.

Who do you look up to?
I look up to any girl or woman who has been pushed down but always got right back up and kept doing what she thinks is right.

What are some ways girls can do STEM activities at home?
You can search for new STEM activities online, create your own projects with things around your house, or improve on a previous project or design.

What advice do you have for other girls?
Remember to be yourself. You can never disappoint anyone by being and showing your true self. Be confident in who you are because one day you will make a difference in our world.

start

Ursula, age 9

What are some things you want us to know about you?
My name means little bear, like Ursa Minor. We call this constellation the Little Dipper, and I can see it in the stars at night. My name is also very special to me because we used to have a dog we called Little Bear (and I love dogs!). I have a mom, dad, and little brother who like to do STEM activities with me, and a dog named Astro who I can pet and play with when I need a break. I like to draw animals and people, and I really love to read a lot. I think reading and drawing are both fun and relaxing, and they exercise my imagination! I also love swimming and playing tennis. I am obsessed with the color pink because it comes in so many different shades! Last summer, I bridged into Juniors with Girl Scouts, and it will be my third year as a Girl Scout.

Where do you do STEM?
I do STEM projects in Girl Scouts, at my house, and sometimes at school. I also do them at STEM Like a Girl workshops, where I get to work with other girls that I've never met, but they're nice and like to do STEM too.

When did you first become interested in STEM?
When I was four, my mom got out an old block and showed it to me and told me it was a cube. She explained the differences between a cube and a square. Then she told me my challenge was to use marshmallows and dried spaghetti to make a cube. It was difficult because the spaghetti kept breaking and I needed all the pieces to be the same size or it wouldn't be a cube. As I got older, I realized that engineering was part of STEM, and I started to get into STEM.

Why do you like doing STEM activities?
I like to do STEM because you can be creative while you do it. I am definitely a creative person. I can do STEM alone or with a group. I like working with a group, though, because I can get a lot of different ideas. And later on, when I do another project, I can use what someone else said to help me make my next project better or stronger.

What is your favorite STEM project you've ever done?
For a school project, I had to design a boat that would neither sink nor leak. It had to be strong enough to hold a die. I used recycled materials like foil, Popsicle sticks, and masking tape, plus a few more things. The first time I tried, my boat was too heavy, and when I put the die in, it leaked and sank, so I had to re-create my design. At first, it took one big change, and then I had to make a small change to it, so it took about three tries to get it to work.

When an experiment or design doesn't turn out how you expected, how does that make you feel?
It makes me feel disappointed, but makes me feel excited, too, because I get to try again. There's no limit to how many times you can try something in

STEM, and you can use what you already know to help you make the best design you can. Later on, that can make you feel good because you might make something that can help people or animals.

What other activities make you feel courageous, confident, and bold?
Reading a book that was difficult or drawing something that was hard for me to learn. Or when I learned to dive, because you have to do that in really deep water, and I had to summon up all my strength to just try. Reading a challenging book helps me remember that I can do really hard things. I remember when I was little and I'd see big, thick books and think, "I will never be able to read those." But a few years later, I was reading those books I thought I could never read.

Who do you look up to?
Marie Curie, the scientist who discovered radioactivity, and Katherine Johnson, who worked with NASA as basically a human computer. They were people who made me think that women can do big things in the world. I also look up to my former second-grade teacher, Brandi, because she helped me get excited about things that I was learning and she was such a nice person. And my mom for encouraging me to do things that I want to do and to not discourage myself when I'm down. She is also my Girl Scout troop leader.

What are some ways girls can do STEM activities at home?
You can visit the Girl Scouts website because they have lots of STEM activities you can try. Cooking is also a great way to think about STEM. You work with a lot of measurements and timing, and it's sort of like engineering because you're making something (even though it's not traditional engineering).

What advice do you have for other girls?
When you're doing STEM, always remember to persevere and be an optimist. And you can learn about STEM and anything else your heart desires in a book, so read a lot!

COLOR SEPARATION

DIFFICULTY LEVEL

What makes up the colors in the markers you use when you draw?

MATERIALS

Safety glasses
White coffee filters (at least 3)
Ruler
Scissors
2 pencils
Set of watercolor markers
2 short, clear cups
2 binder clips
1 permanent marker
Water
Isopropyl alcohol (rubbing alcohol)

TEST #1: COLOR SEPARATION

METHOD

1. Put on your safety glasses. When working with chemicals, it is always important to protect your eyes!

2. Cut a coffee filter into a rectangle approximately 2 in. wide. The length will depend on how tall your cup is. The rectangle should be slightly longer than the depth of your cup.

3. With your pencil, draw 4 small circles in a row, approximately 1/2 in. away from one of the short ends of the rectangle.

STEP 2

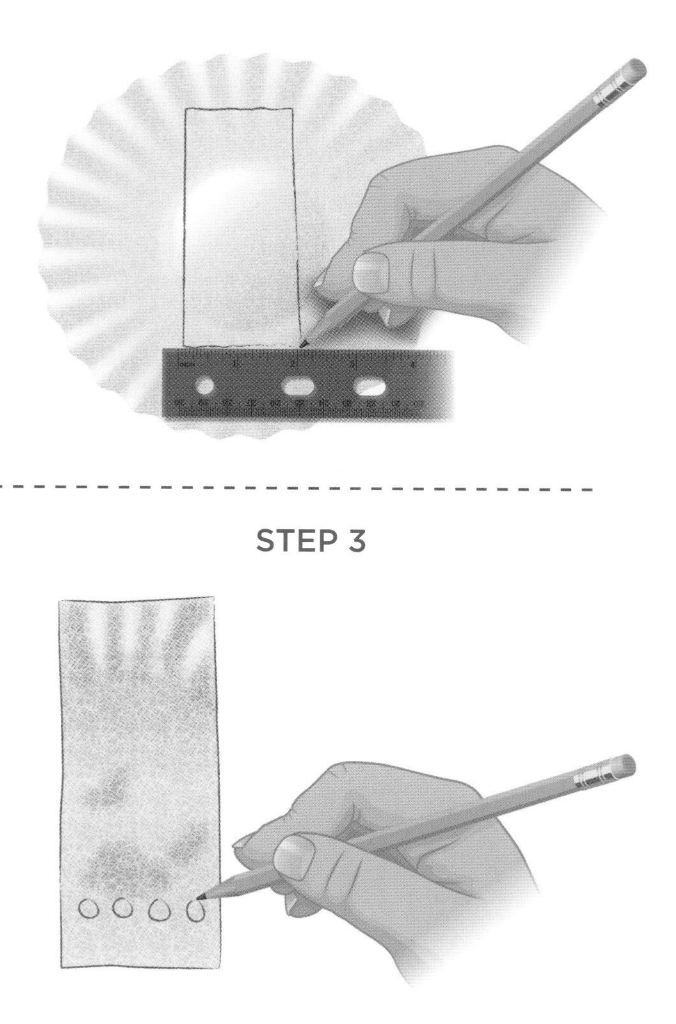

4. Color in each circle with one of four different watercolor markers. Red, orange, green, and black work well, but experiment with both primary and secondary colors.

Roll the other end of the rectangle around a pencil, and secure it with a binder clip.

5. Lower the filter into one of the cups, circles-side down, so that it hovers just above the bottom of the cup. You don't want the end of the filter to touch the bottom of the cup. If necessary, adjust the length of the filter by rolling it farther around the pencil and resecuring the binder clip.

STEP 4

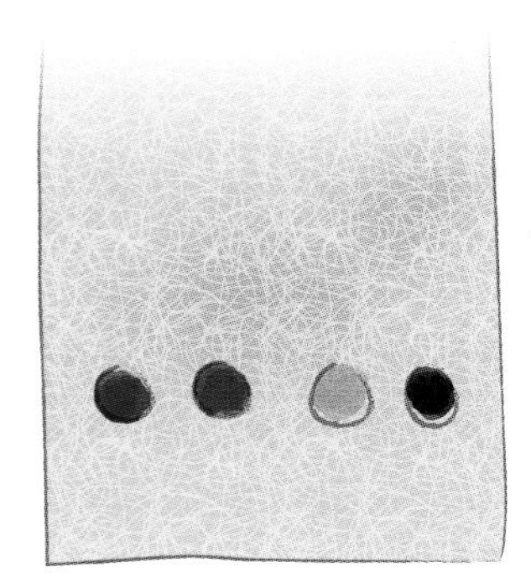

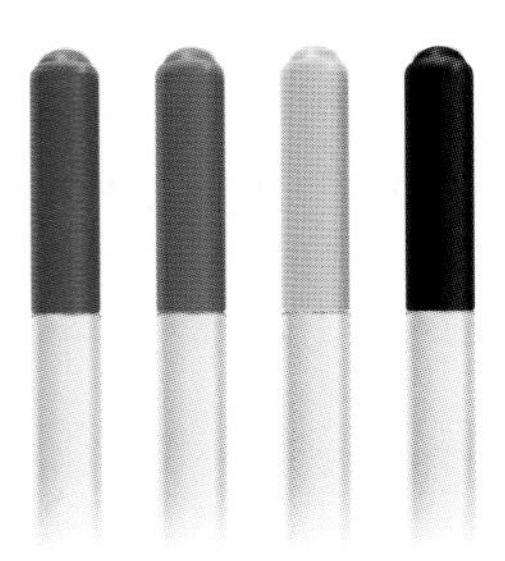

6. Slowly pour water (the solvent) into the bottom of the cup so that it just covers the bottom of the filter paper, but not deep enough to touch the colored circles. Be careful not to pour the water above the circles or along the length of the filter paper.

7. Start your observations. The water will slowly be absorbed upward along the filter. What happens to the colors as the water moves up the filter? Do the primary, secondary, and black or brown colors react differently? Record your observations on your experiment test sheet.

8. After several minutes, the colors will stop moving up the coffee filter. You can remove it from the cup and allow it to dry.

STEM APPLICATION

Have you ever thought about what makes markers different colors? The various colors come from different *molecules*, or chemicals that make up the ink. Each color has its own unique combination of ink molecules. Molecules differ in size and weight. Some are heavier than others, so they behave differently. Scientists can separate different molecules using a method called *paper chromatography*.

The molecules (in this experiment, the molecules that make up the ink from the markers) are placed on a piece of filter paper and exposed to a *solvent*, a liquid that dissolves the molecules. As the solvent (in this experiment, the water) is absorbed upward by the filter paper, it takes the soluble part of the ink molecules with it. These color molecules move up the paper at different rates. When this happens, the different colors in the ink become visible. How far up the paper the molecules travel is related to their size. Smaller particles move farther and faster than bigger ones.

Primary colors (red, blue, and yellow) are made of only one color molecule, so you only see that color on the filter paper. Secondary colors (green, purple, and orange) are made from two color molecules, which become separated as they travel up the paper at different rates. For example, if you started with a green marker, you should see it separate into yellow and blue along the filter paper. The colors black and brown are made up of several colors, which is why you may see shades of red, green, and yellow on the filter paper.

TEST #2: EFFECT OF SOLVENT CHOICE

METHOD

1. Put on your safety glasses. When working with chemicals, it is always important to protect your eyes!

2. Cut a coffee filter into a rectangle approximately 2 in. wide. The length will depend on how tall your cup is. It should be slightly longer than the depth of your cup. Repeat with a second coffee filter.

3. With your pencil, draw two small circles side by side on each filter, located approximately 1/2 in. away from one of the short ends of the rectangle. Label one rectangle "water" and the other "alcohol." You will be comparing the two solvents.

4. On each rectangle, color one circle with a watercolor marker and the other with a permanent marker.

STEP 3

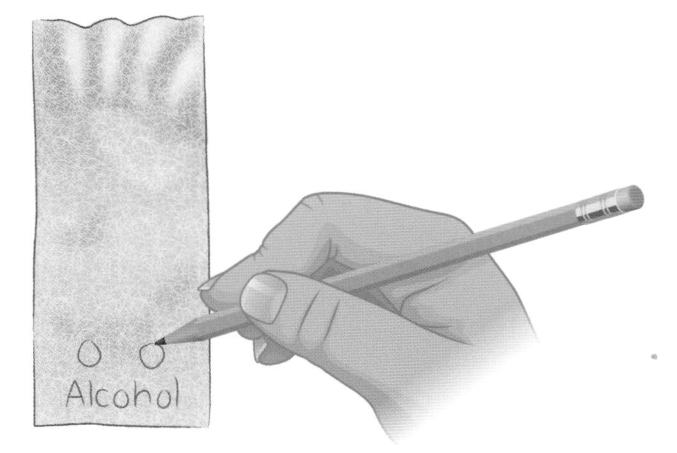

5. Lower each filter into a separate glass and secure it in place like you did in Test #1. Check that the end of the filter is just above the bottom of the cup, not touching it. If necessary, adjust the length of the filter, and resecure the binder clip.

6. Slowly pour water into the bottom of one cup and isopropyl alcohol into the bottom of the other cup so that the liquid just covers the bottom of the filter paper. As before, be careful not to pour the solvent above the colored circles or along the length of the filter paper.

STEP 6

7. Start your observations. How do the different types of markers behave in the two different solvents? Is there a difference between water and isopropyl alcohol as solvents? Record your observations on your experiment test sheet. Note: The alcohol takes much longer than the water to travel up the filter.

8. After several minutes, the colors will stop moving up the coffee filters. You can remove them from the glass and allow them to dry.

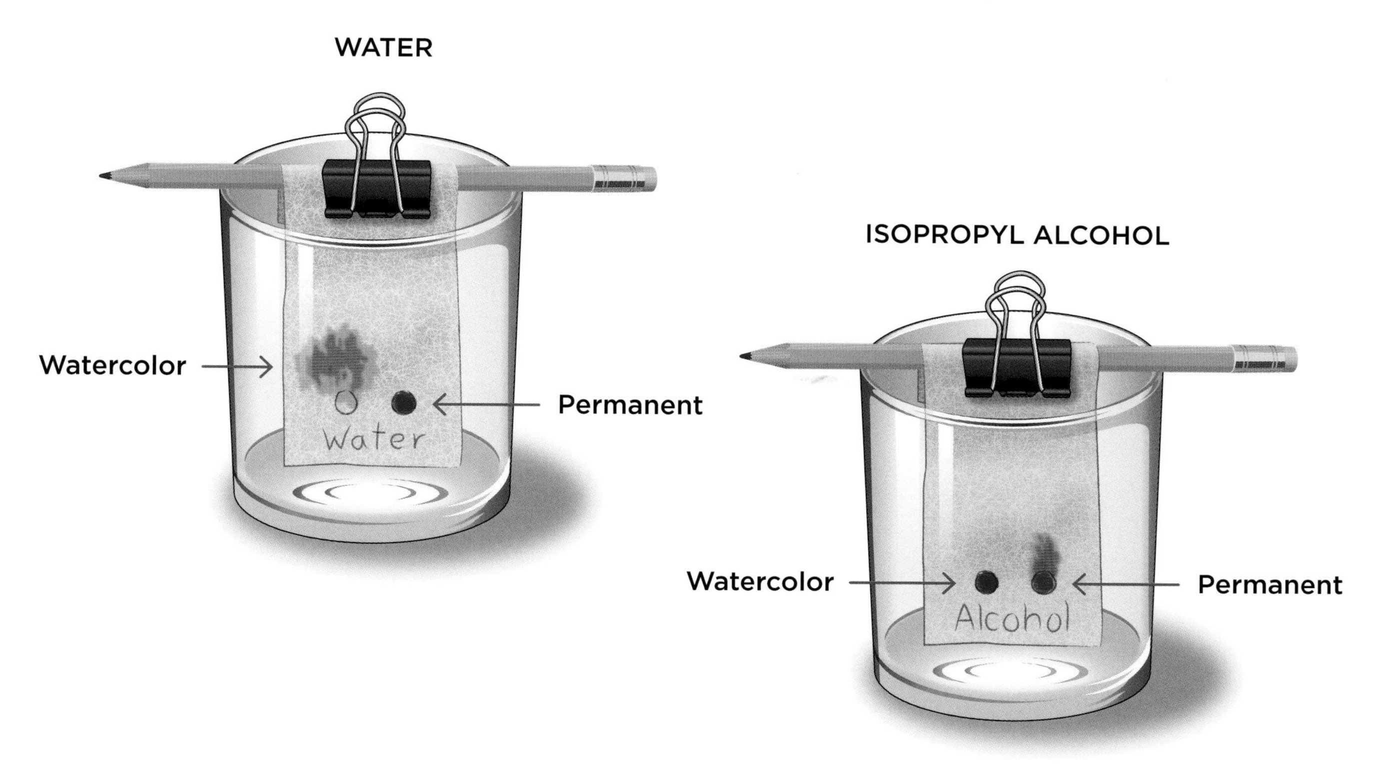

STEM APPLICATION

In order for the color molecules in the marker ink to move up the filter paper, they must be able to dissolve in the solvent. Ink from a watercolor marker is designed to dissolve in water, while ink from a permanent marker will not. This is why you can easily wash watercolor markers off your skin or clothes, but permanent markers are much harder to remove. When the two types of ink are exposed to water as a solvent, the watercolor molecules will be pulled up the filter paper by the water, but the permanent marker molecules will remain in the original spot. On the other hand, permanent marker ink will dissolve in a different solvent, like alcohol. In this case, the permanent marker molecules are pulled up the filter paper by the isopropyl alcohol, but the watercolor marker molecules remain in the original spot.

Kate, age 11

What are some things you want us to know about you?
I am a twin and I love the arts. I am an avid reader and I enjoy cooking. I value spending time with friends and family. One of my favorite pastimes is creating anything and everything!

Where do you do STEM?
I participate in STEM activities in Girl Scouts and at my school.

When did you first become interested in STEM?
I first became interested in STEM when I was in third grade. I attended a summer STEM camp and loved it. We got to try a tiny bit of a ton of different types of STEM.

Why do you like doing STEM activities?
I love participating in STEM because you get to be creative. Whenever I do STEM, I always have a blast with my friends and learn a ton.

What is your favorite STEM project you've ever done?
My favorite project I've done is when my twin sister and I and one of my friends got together and learned about chemistry and different types of acids. We tried to do a "pop-a-bag" experiment with no instructions. We got to try and figure out what materials would make the bag pop fastest. It was awesome!

When an experiment or design doesn't turn out how you expected, how does that make you feel?
When I make a mistake or something surprises me, I'm happy. I was wrong, which means I get to try again and spend more time learning and engineering.

What other activities make you feel courageous, confident, and bold?
I enjoy writing and being onstage. I also read a ton and am a very active Girl Scout. Both activities allow me to be creative and learn, just like I do in STEM.

Who do you look up to?
I look up to both my parents because they always look out for me and are all-around awesome. I also look up to my grandparents because they've supported me in everything I've done. And finally, my babysitter, Haley, and my teacher, Ms. Norman, because they both taught me to be creative and think outside the box.

What are some ways girls can do STEM activities at home?
If you can't get together with friends in person, connecting online and completing experiments virtually is a fun way to work together. It is a little more difficult, but I really enjoy it.

What advice do you have for other girls?
I always think of a saying I heard when I do STEM. The saying is "Science is today's tomorrow." So when you participate in STEM activities, remember that you can make a difference in tomorrow, and keep working. One day you might just be a famous scientist!

Sakina, age 11

What are some things you want us to know about you?
I like to draw and swim. My favorite things to draw are animals, flowers, and different animations. I also like to edit photos. One of my favorite things to do is to take pictures of celebrities, myself, or other people and edit them in fun ways.

Where do you do STEM?
I started doing STEM at my school and then signed up for a coding camp at my local library.

When did you first become interested in STEM?
I started liking coding when I was introduced to the computer program Scratch. Scratch is basically a way to animate characters and tell them what to do to make a story.

Why do you like doing STEM activities?
I really enjoyed Scratch because it allows you to make the animations you draw do stuff for you. It was fun!

What is your favorite STEM project you've ever done?
I am still working on my first project, but I'm making and editing a video of styling my hair.

When an experiment or design doesn't turn out how you expected, how does that make you feel?
I feel sad at first, but then when I get ideas, it makes me motivated and happy.

What other activities make you feel courageous, confident, and bold?
It makes me confident when I can make someone laugh. I sometimes make funny videos to share with my friends (approved by my parents, of course). Jumping into a ten-foot swimming pool made me feel courageous.

Who do you look up to?
I admire my mother because she always encourages and supports me.

What are some ways girls can do STEM activities at home?
You can go online and learn how to create stuff. Topics like "how to braid" or "how to edit and put an effect on a picture" will help you get ideas. You can also try out Scratch (it's free!) and make your own animations.

What advice do you have for other girls?
Do what you want to do and be true to yourself!

Juliette, age 11

What are some things you want us to know about you?
I have two dogs. One of my dogs is named Sapphire, and we call her the "golden dingo." Anna is our rescue dog. She loves snuggles and is probably the sweetest dog. When we say, "Walkies!" they both go crazy because they love the outdoors. I am a musician, primarily a drummer, but I can play ukulele and a little piano. I'm crazy about graphic novels, stargazing, traveling (I hope to visit at least half the world someday), and I've been doing jiujitsu since I was five years old.

Where do you do STEM?
At Girl Scouts, at STEM Like a Girl, and at home. I do a lot of STEM activities at home with my parents and a few friends. When my mom was our Girl Scout leader, she had us do a lot of different and fun activities, and that got me into other group STEM activities.

When did you first become interested in STEM?
I've been doing STEM activities with my family since I was, like, three years old but didn't realize I really liked STEM stuff until I started doing things with my Girl Scout troop. I think that's when I first learned that STEM was a thing!

Why do you like doing STEM activities?
I love doing STEM activities because there is always something new to learn. They introduce me to different ways of learning and new ways of thinking about problems. I really love chemistry! I love to mix ingredients and do chemistry—especially baking—and to learn about the chemicals, and why they react the way they do. I love to keep notes on my experiments so I can look back and see reactions and the differences between the chemicals.

What is your favorite STEM project you've ever done?
There are so many! I really loved the DNA project with STEM Like a Girl because I got to see a version of "me" that I'd never seen before.

When an experiment or design doesn't turn out how you expected, how does that make you feel?
When an experiment doesn't turn out, I get curious about what made it not work. I wonder, was it the materials I used? Was it the way I built it? Or was it something else that wasn't working, like a battery that was out or a wire that short-circuited? I keep trying to figure it out and don't stop until I do, because I'm persistent.

What other activities make you feel courageous, confident, and bold?
I'm not very comfortable standing out in a crowd, but my drum teacher taught me that if something doesn't feel or sound right, to change it so it fits or so it feels right. I use this trick in almost everything I do, and I think it helps me feel more courageous.

SPACE LIFE
OUTTA

Who do you look up to?
I look up to my parents most of all because they think differently from me and they help me think outside the box. They are really creative, and whenever something doesn't work and I'm confused or curious, they always help me figure it out.

What are some ways girls can do STEM activities at home?
Some of my favorite STEM activities to do at home use everyday supplies, like soap, candy, spoons, cardboard boxes, tape, pencils, and paper. I love to make different kinds of slimes and putty using different ingredients. And I love, love, love to bake, especially when I get to work with different types of leavening agents (like baking soda, baking powder, and yeast). We always seem to have some science kits around our house too.

What advice do you have for other girls?
I think it's really important for girls to know that we are capable of doing many things, and that even when it is hard, we have to persevere and keep pushing forward.

FIZZY FLOWERS DIFFICULTY LEVEL

Make your own bath bomb and discover what makes it "fizz" in water and what makes it smell so nice. (Hint: The answer is *chemistry*.)

MATERIALS

Safety glasses
Baking soda
Citric acid
Epsom salt
Cornstarch
Olive oil (or other liquid oil of your choice)
Water
Essential oil (optional)
Measuring cups and spoons
Large bowl
Spoon
Small jar with a lid
Silicone flower molds or small disposable plastic cups

METHOD

1. Put on your safety glasses. When working with chemicals, it is always important to protect your eyes!

2. Mix together the dry ingredients. In a large bowl, use a spoon to combine:

1/2 cup baking soda
1/4 cup citric acid
1/4 cup Epsom salt
1/4 cup cornstarch

(Note: Although it is completely safe, citric acid can damage some countertops, including marble and granite, so you may want to put a place mat under your work area. Clean up any spills right away.)

3. Next, mix together the liquid ingredients. In a small jar with a lid, combine:

2 tablespoons oil
1 1/2 teaspoons water
Several drops of essential oil (optional)

Shake well. It is very important that the liquid ingredients be well mixed, so if they sit for a while, give them another good shake since the oil and water may start to separate.

4. Slowly pour the liquid ingredients into the dry ingredients, stirring with a spoon. Make observations about what you see and hear and record them on your experiment test sheet. That fizz you hear is the acid-base reaction starting!

5. The mixture should be the consistency of crumbly sand. You will be tempted to add more liquid, but don't! Use your spoon to break apart any large pieces.

6. Scoop some of the mixture into each mold or cup. Press down with the back of a spoon or your fingers to pack it well. You want to make sure it's packed tightly into your mold or cup so that it doesn't fall apart when you remove the dry bath bomb. Alternatively, you can form the mixture into balls by pressing it together and shaping it with your hands.

CE LIFE

7. Let the bath bombs dry overnight before removing them from the mold. Store the dried bath bombs in a zipper-sealed plastic bag or airtight container.

8. When you're ready for a bath, drop one or two of the bath bombs into the running water. Watch and listen as the acid-base reaction continues!

STEP 7

STEP 8

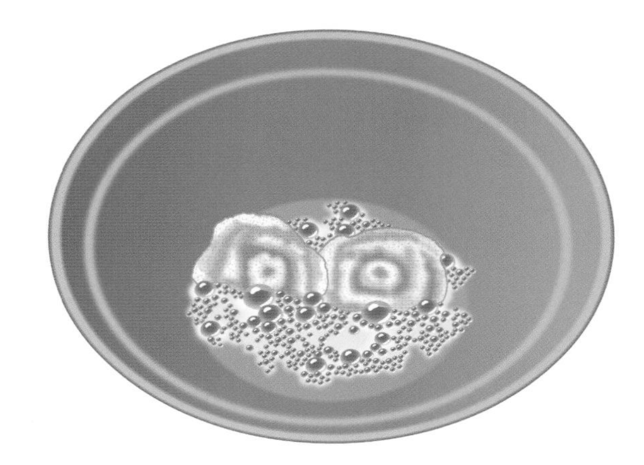

STEM APPLICATION

What makes a bath bomb fizz? This reaction is known as *acid-base chemistry*. Baking soda, or sodium bicarbonate, is a base, which means it has a pH higher than 7. Citric acid is an acid, which means its pH is lower than 7. When an acid is mixed with a base, a reaction occurs as the two chemicals try to neutralize each other (that is, to get to a pH of 7). One more chemical is needed to initiate this reaction: water! The dry baking soda and citric acid don't react with each other until they become wet. When you combined the liquid ingredients with the dry ingredients, you probably observed a small fizz. This meant the reaction was starting, but there wasn't enough water to push it to completion. You need to limit the amount of water when forming the bath bomb so that the big fizz is saved for when you drop it into the bath. In this experiment, the acid-base reaction produces carbon dioxide gas, which makes the fizzy bubbles.

Have you ever thought about what makes a bath bomb smell nice? You can thank chemistry for that too! The scents in bath products are made of small molecules that are separated, isolated, and identified by a chemical process called *liquid chromatography*. Some scents are made from just one molecule, while others are a combination of several molecules. For example, the smell of lavender is a combination of four small molecules, and wild orange is a single molecule.

LAVENDER

(R)-Linalool

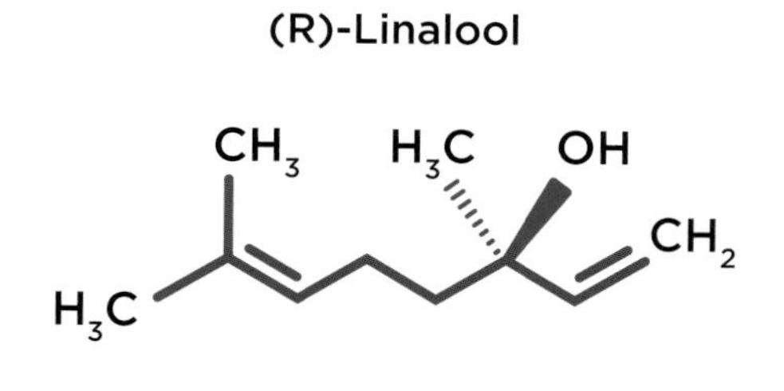

Linalyl Acetate

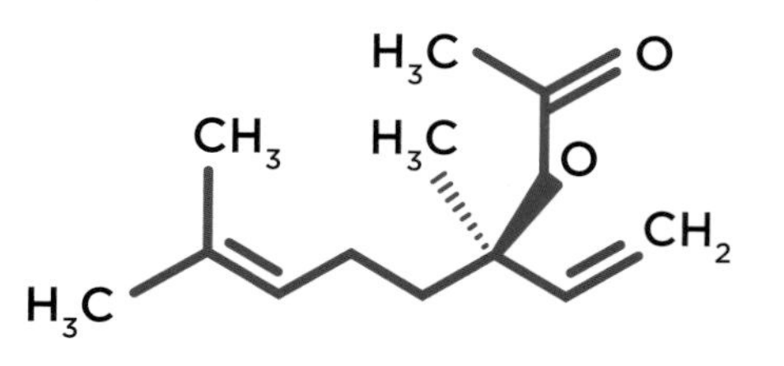

(R)-Lavandulol

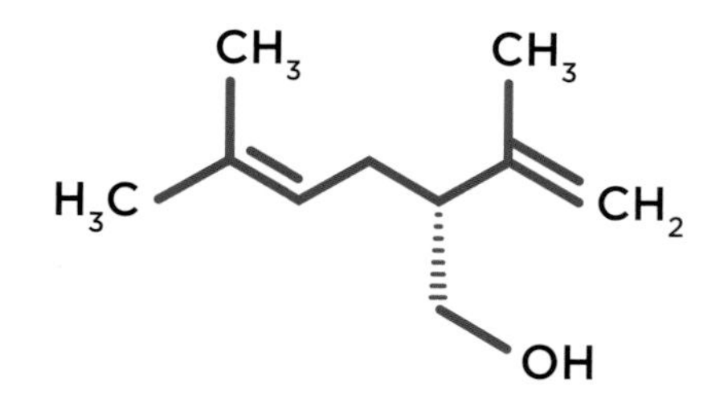

Lavandulyl Acetate

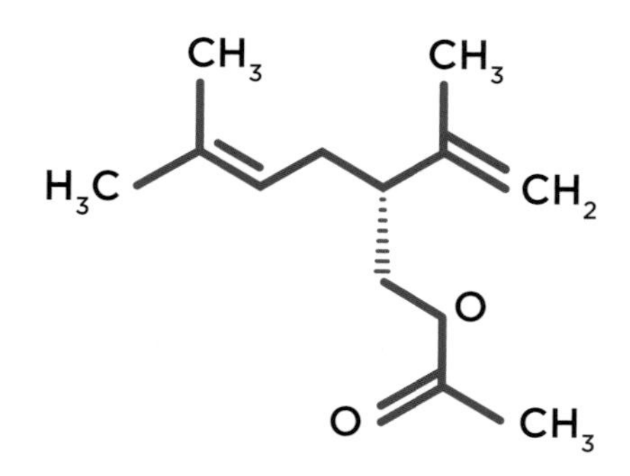

WILD ORANGE

Limonene

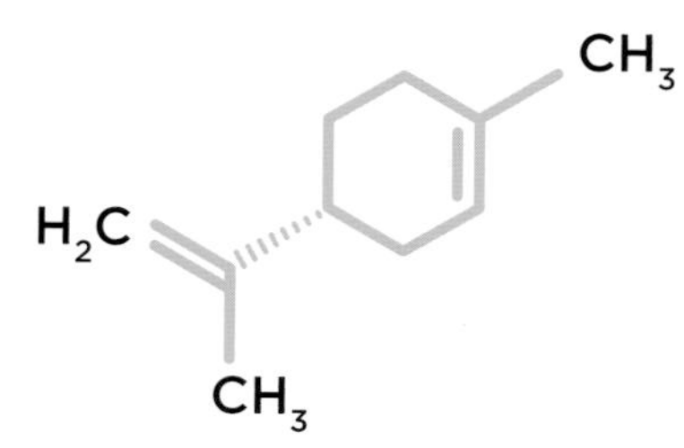

Olivia, age 11

What are some things you want us to know about you?
Some things to know about me are that I am a bookworm and love school. I love everything space-related, and I love to do crafts and to build gadgets from items I can find around my house. Aside from STEM, I am passionate about the arts, including piano and drawing. In addition, I like to stay active by taking tennis lessons, swimming, and martial arts.

Where do you do STEM?
I do STEM almost anywhere, depending on what I need to do. I mainly do STEM at school in STEM class and during my after-school programs. However, I love to do little fun projects at home, too, such as building a long grabber made of markers and paper towel rolls to reach items behind the couch.

When did you first become interested in STEM?
I have been interested in STEM since even before I learned there was a term for it. For instance, I started building Legos when I was three years old. I would then use my creativity to build other things out of broken Legos, like houses and secret passageways. As I learned more about STEM in school, I found that I had a strong passion for the subjects related to science, technology, and math.

Why do you like doing STEM activities?
I like doing STEM activities because they test not only your brain but also your patience. They are fun and somewhat addicting (in a good way). You think you have it solved, then it breaks or falls. That usually earns a few laughs. Then we try again, and again, and again, until we get it to work correctly! It provides self-satisfaction when you successfully complete a STEM challenge. It encourages you to seek out other STEM challenges.

What is your favorite STEM project you've ever done?
My favorite STEM project I have ever done would have to be the chip circle. Last year during a summer program, we were challenged to build a circle out of Pringles chips, but there was a catch: We could not have it in 2D. It had to stand upright in a wheel shape. If we completed the project, we would be able to eat the Pringles! My group kept trying and failed every time. Many tries and cracked Pringles later, we completed the project! Then we slowly ate apart the Pringles wheel, top to bottom.

When an experiment or design doesn't turn out how you expected, how does that make you feel?
When an experiment or design does not turn out how I expected, I usually feel a bit disappointed that it did not work, but it motivates me to find another way to make it work. I am not discouraged to take it apart and try again.

What other activities make you feel courageous, confident, and bold?
Some other activities that make me feel courageous, confident, and bold are rock climbing, doing math problems, reading, and drawing. I approach each of these as challenges to solve or overcome.

Who do you look up to?
I mainly look up to my parents because they are very hardworking and love me unconditionally. I also look up to Neil deGrasse Tyson, a famous astrophysicist, because he is not only a genius, but also makes complex things simple to understand.

What are some ways girls can do STEM activities at home?
Girls can do STEM activities at home by gathering up household materials they might have lying around and searching on the web for cool STEM activities to try. They could also use their imagination to come up with something they would like to create and use household materials to build it.

What advice do you have for other girls?
Never give up! That is very important. Perseverance is a must-have when doing STEM challenges, but also for getting through life. Secondly, it is important to be flexible and open-minded because ideas can be changed or improved. A third piece of advice would be to work hard. Working hard will lead to a higher chance of success in everything you do. Finally, be creative. Thinking outside the box will work well when solving puzzles, problems, and challenges.

Laila, age 9

What are some things you want us to know about you?
I enjoy building stuff and making new friends. I like to build handyman projects like a laptop out of a pizza box. I also built a cart that you can store food on. I want to be an astronaut. It would be fun to be an astronaut as you are floating around and going to planets.

Where do you do STEM?
I attend an after-school program called Circuit and joined a coding club at my local library.

When did you first become interested in STEM?
I was always interested because I think science is cool. I think dissecting animals is an extremely cool thing to do and watching the life cycles of chicks, butterflies, and mealworms.

Why do you like doing STEM activities?
I like doing different STEM activities because it is fun how you can make things work. I once made eyeglasses out of plastic spoons and forks.

What is your favorite STEM project you've ever done?
In coding club, I did an animation on endangered animals. It was quite interesting. I did it on turtles because I thought it was fascinating how turtles breathe from their butts (hehe)!

When an experiment or design doesn't turn out how you expected, how does that make you feel?
If something goes wrong, I usually start over. I feel sad because I worked so hard on it and that one mistake made me start all over.

What other activities make you feel courageous, confident, and bold?
Holding a baby crocodile made me not scared anymore. Also, when I did rock climbing for the first time and reached the top, I felt very courageous.

Who do you look up to?
I look up to my daddy because he always tells the truth and everybody goes to him for his advice.

What are some ways girls can do STEM activities at home?
You can always go online and search for activities. I usually type "fun STEM activities at home" and get lots of ideas to try.

What advice do you have for other girls?
Do what makes you happy and be kind to one another!

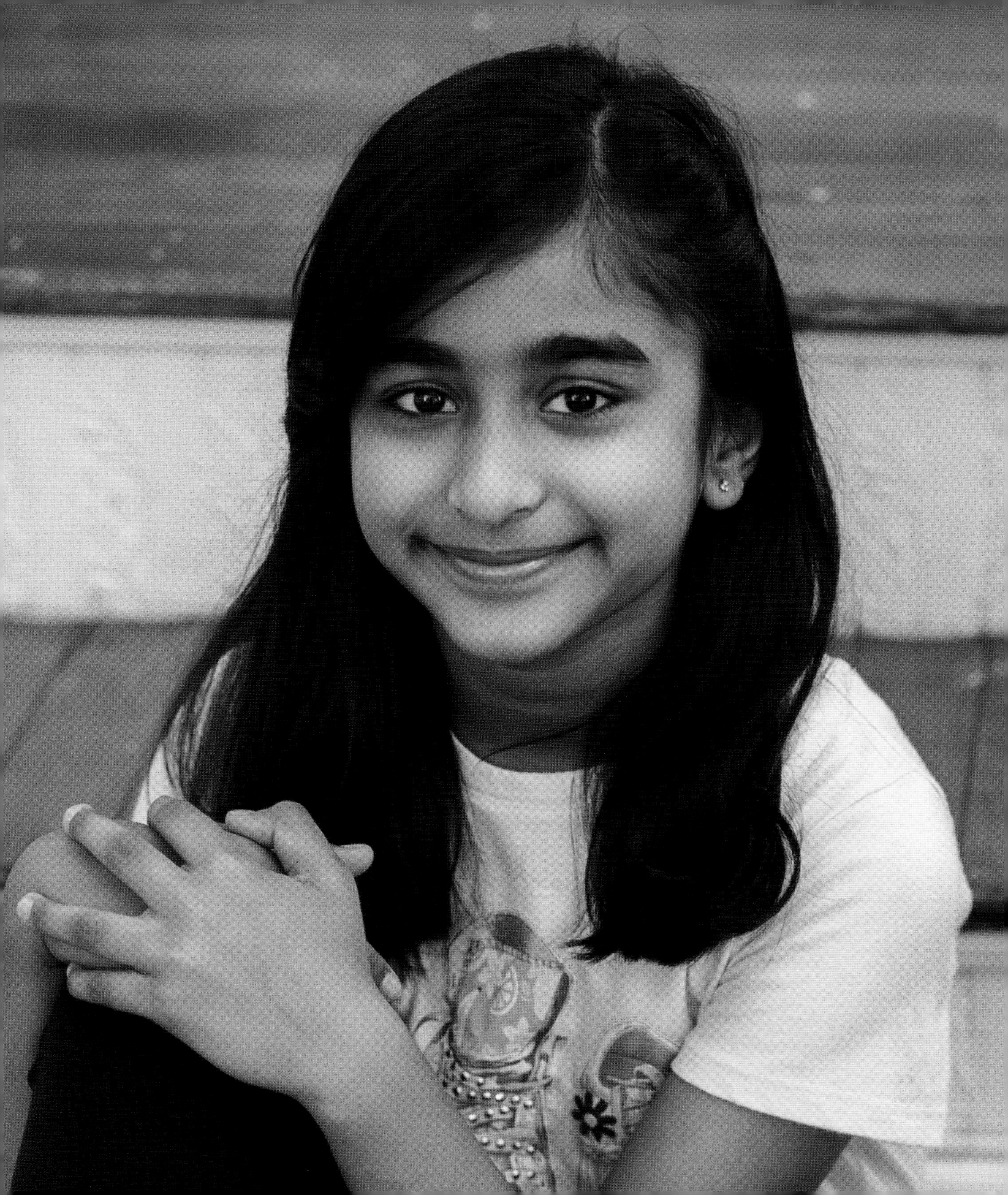

MARBLE MAZE DIFFICULTY LEVEL

Design and build a maze to successfully move a marble from start to finish without the marble falling off the plate.

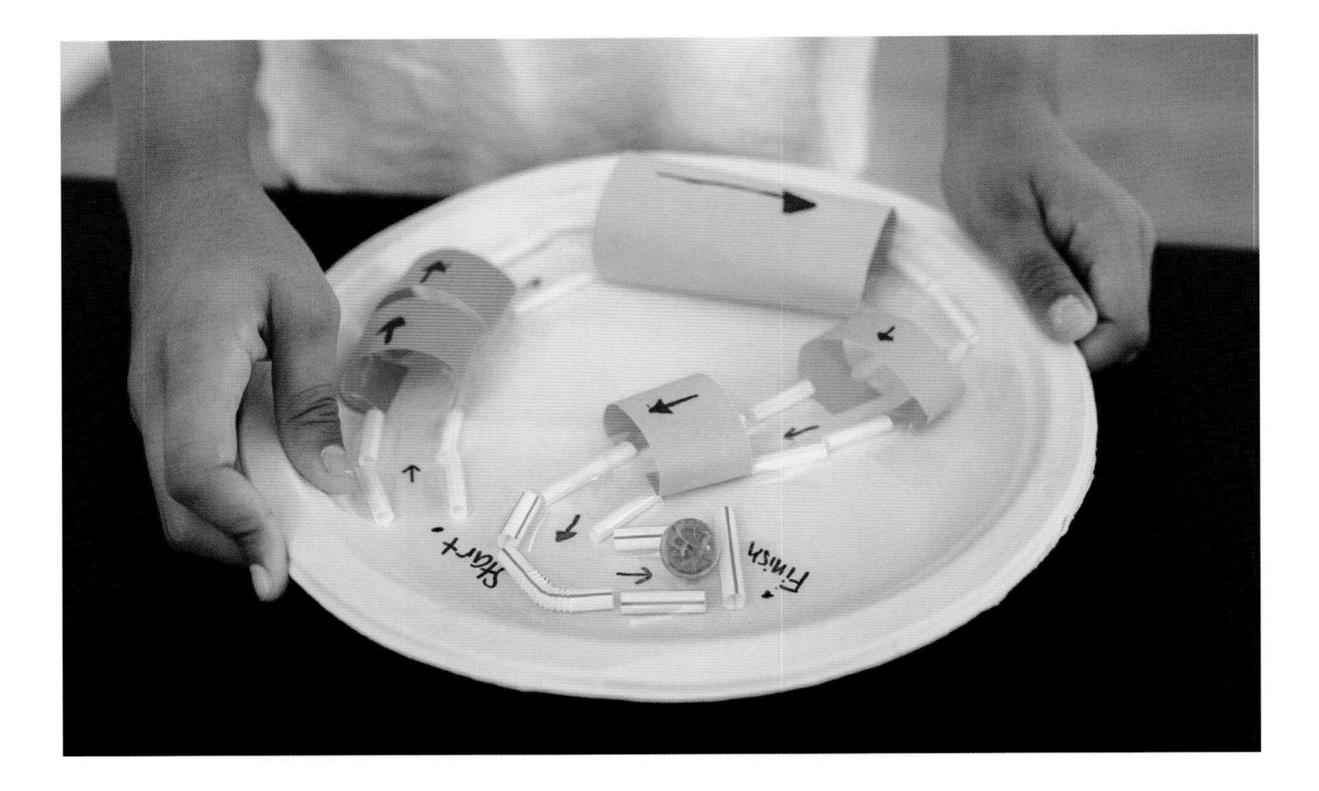

MATERIALS

- Heavy-duty paper plate
- Marker
- Several straws
- Card stock, construction paper, or craft foam
- Scissors
- Clear tape
- 1 marble

METHOD

1. Determine where the start and finish points for your maze will be. Use the marker to label these points on the plate.

STEP 1

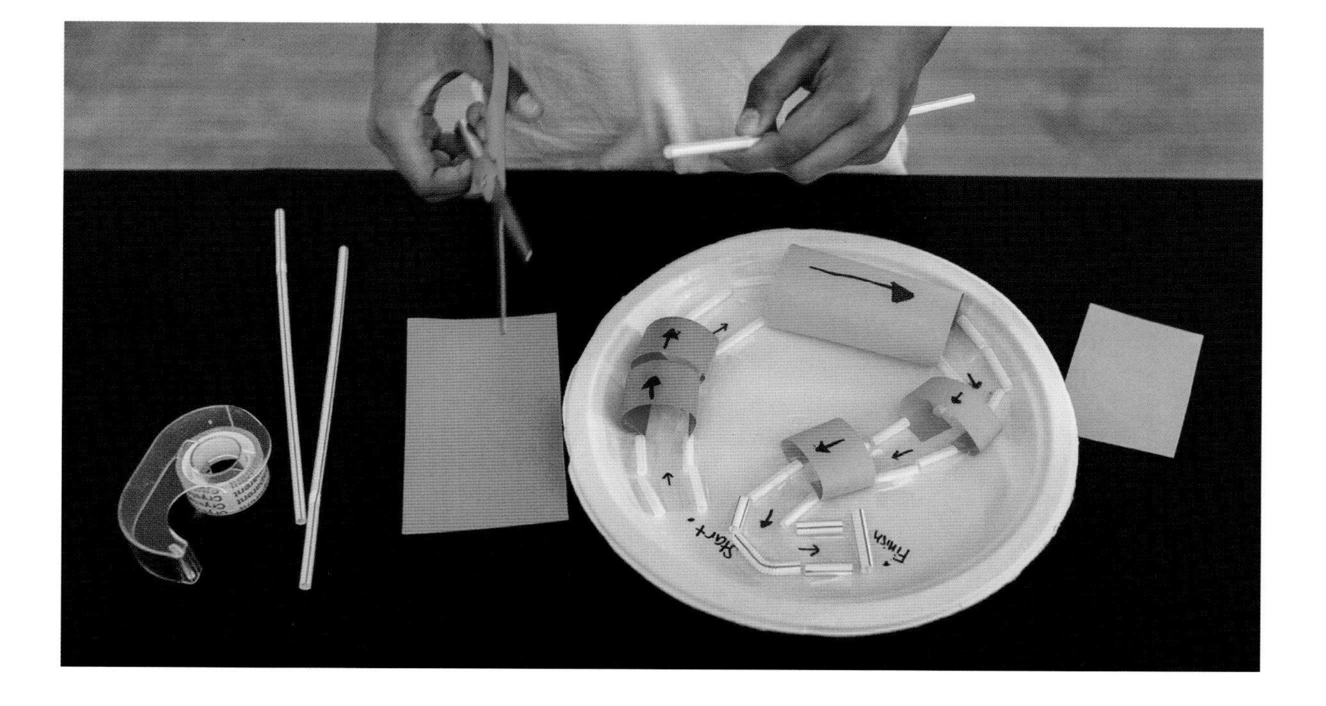

2. Using straws and paper or craft foam, design a maze that includes the following features:

- At least 2 arches or tunnels, using construction paper, craft foam, etc.
- At least 1 channel, using parallel pieces of straws
- Arrows showing the direction of movement

Use the scissors to cut shapes and tape them to the plate to create a path for your marble. There is no right or wrong way to design your maze, so get creative!

3. Test your design using a marble and make observations. What did you learn from each test run? What improvements can you make? How did you feel if your design didn't work the way you thought it would? Record your observations on your experiment test sheet.

4. If your marble fell off the plate or couldn't complete the maze, make changes to your design and test again.

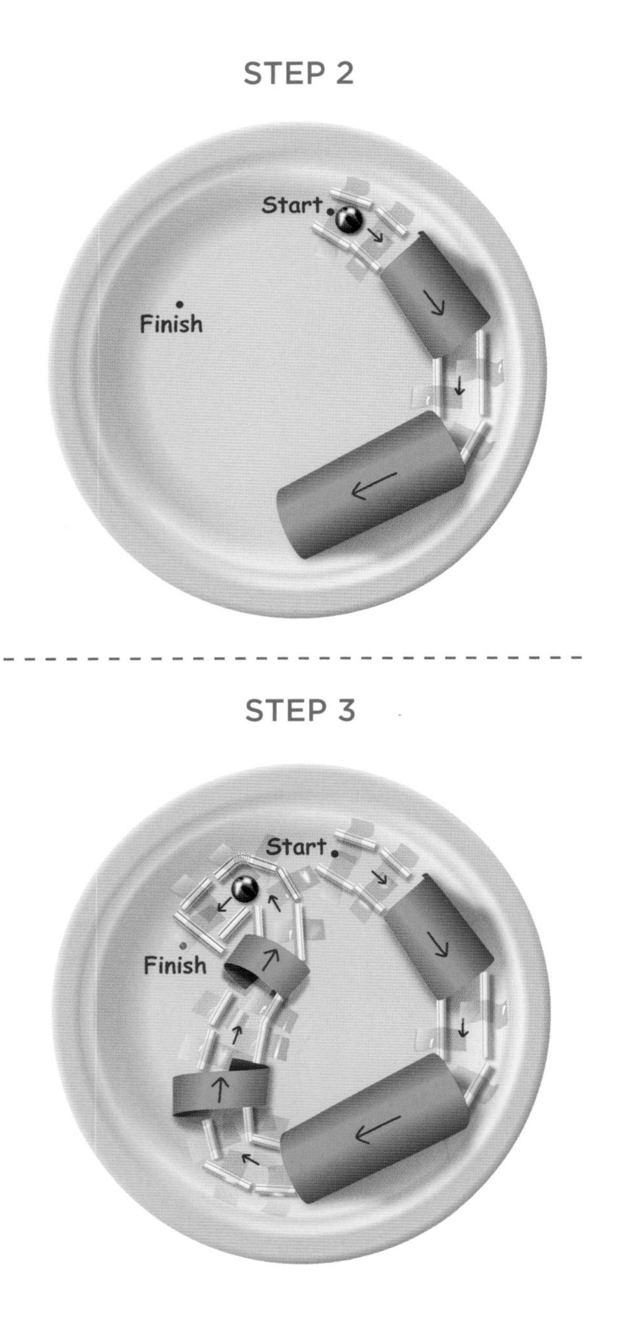

STEM APPLICATION

Engineers need to consider many different design elements when planning out a road or pathway. How are cars, pedestrians, bikers, or trains going to safely navigate from their starting point to their destination? Sometimes the path is straightforward, but other times engineers need to work around geographic obstacles (like mountains or waterways), man-made structures (like buildings and existing roadways), or design requirements (like building-code restrictions, zoning laws, and accessibility).

K1000

Rachel, age 10

What are some things you want us to know about you?
I do photography. I am learning to knit, and I love designing clothes. I am a very creative person. I love to plan things. I make friends easily and love to play outside. I am interested in woodworking.

Where do you do STEM?
I do STEM at home, at Girl Scouts, at my local library, while camping, and on vacation.

When did you first become interested in STEM?
I first became interested in STEM when I was seven. The four reasons I became interested in it are: (1) my grandpa is an engineer and I like to talk to him, (2) I ask lots of questions and am very creative, (3) I enjoy doing experiments, and (4) my mom taught me how to use technology and Photoshop.

Why do you like doing STEM activities?
I like doing STEM activities because it makes me question the world, and the activities are a lot of fun to me.

What is your favorite STEM project you've ever done?
My favorite STEM project I've ever done is either when I made my own ninth birthday cake and fondant from scratch. (I used math because I had to measure and I had to double and triple recipes.) Or it's when my grandpa came to my Girl Scout meeting and did an engineering lesson and helped us do chemistry experiments so we could earn our STEM badges.

When an experiment or design doesn't turn out how you expected, how does that make you feel?
At first, it makes me want give up, but then I get determined and work even harder to do better and improve it.

What other activities make you feel courageous, confident, and bold?
Other activities that make me feel confident, courageous, and bold are swimming, baking and cooking, reading, photography, camping, sewing and designing clothes, and acing tests.

Who do you look up to?
I look up to many people. My parents, who successfully run their own businesses (architecture and photography); my grandma, who is a retired nurse; Anne Frank, who wrote about her life in the secret annex; and Marie Curie, who was the first woman to win a Nobel Prize and ran life-changing research on radioactivity.

What are some ways girls can do STEM activities at home?
Some ways girls can do STEM activities at home are baking, learning photography, woodworking, coding, and doing science experiments.

What advice do you have for other girls?
Be yourself, and girls can do anything. All in all, just explore the world and have fun. I mostly would advise you to stand out from the rest of the crowd and have fun when you do. Also, never be self-conscious about how you look. I make that mistake all the time.

Birtukan, age 10

What are some things you want us to know about you?
I love dogs and I'm in fifth grade. I like getting creative with things even when it doesn't always go as planned. I just try it again or move on to a different thing. Also, I'm awesome!

Where do you do STEM?
I participate in Girl Scouts, where we do a lot of creative STEM things, and I did an art and STEM camp when I was younger. I also love to build things in my grandpa's studio or do STEM kits and projects at home.

When did you first become interested in STEM?
I went to a STEM Like a Girl event, and it was really fun. I got to really think about stuff. I also like doing science class and math class at school. I like building things, and I just learned to weld a pot for the first time.

Why do you like doing STEM activities?
I like doing math because it challenges my brain a lot. Engineering is really fun, and you can get creative with it. I really enjoy science and like coding with Scratch.

What is your favorite STEM project you've ever done?
I built an upcycled fairy house using things that were probably going to get thrown away. I collected items and built a house. It was really cool even though my dog pulled it apart!

When an experiment or design doesn't turn out how you expected, how does that make you feel?
It makes me feel a little upset and a bit bummed that it didn't work out. But then I calm down and I try it again. Sometimes something goes wrong, but if you try it again, it goes right, so never give up after the first try.

What other activities make you feel courageous, confident, and bold?
Earning badges at Girls Scouts makes me feel strong. Doing art and music makes me feel confident.

Who do you look up to?
A lot of members of my family are scientists or artists, and I look up to them, including my mom. She's a scientist and is pretty cool. I look up to my grandpa, who is an artist, and my other grandpa, who is ninety-nine and is a doctor.

What are some ways girls can do STEM activities at home?
You can find things around your house to build with or ask a parent to write down math equations and challenge you to solve them. You can do at-home science experiments, but just make sure they don't blow up your house! There are also fun engineering kits you can order where everything you need to build the projects is included. There are great websites that can teach you how to code.

What advice do you have for other girls?
If things don't go your way, never give up! Always believe in yourself.

LEAVE
NO ONE
BEHIND

Zahra, age 11

What are some things you want us to know about you?
I love cookies, stuffed toys, and noodles. My favorite ice cream flavor is Monster Cookie. I like making comics and playing video games. One goal by the end of this year is to finally do a handstand. I love the superhero Ms. Marvel because she is stretchy, she is strategic, and she actually knows what she is doing.

Where do you do STEM?
I do STEM activities at home, at school, in online classes, and at events like STEM Like a Girl. I also do STEM at our science museum.

When did you first become interested in STEM?
When I was in kindergarten, my brother (who is two years older than me) and I made "potions" in our yard by mixing random leaves, flowers, and plants together and adding water. I remember we filled little containers with that potion and observed how it changed over a few days. Let me just say, it transformed from a nice, fragrant mixture to a terrible-smelling concoction.

Why do you like doing STEM activities?
They're fun! Every time you try, you learn something new and make discoveries. You can also experiment with really cool things. At the science museum in my town, I can make colored fire using different elements or design the structural plan of a coastal city to save it from a tsunami.

What is your favorite STEM project you've ever done?
Probably when I extracted and put my own DNA in a necklace, and I don't mean I just put a hair in it. You could actually see strands of DNA! My second favorite is when I used a magnifying glass and the sun to burn a piece of paper. I colored the spot where I was going to burn with a black Sharpie to make it easier to burn.

When an experiment or design doesn't turn out how you expected, how does that make you feel?
Sad at first, but then happy if I can make it better, and even happier when I try it again and it works.

What other activities make you feel courageous, confident, and bold?
Fighting (taekwondo), playing freeze tag, building stuff in Minecraft. Also, reading and competing in Battle of the Books with my school team. We even won the state championship.

Who do you look up to?
No one, except maybe my future self, because I will be working hard to make this world a better place to live. I imagine myself being smarter, stronger, more kind and fierce.

What are some ways girls can do STEM activities at home?
They can experiment with things they want to know more about. Being curious, wondering about things, and finding information you didn't know is a great way to learn more about any topic in STEM.

What advice do you have for other girls?
Never listen to anyone who tells you to NOT eat cookies. Don't be afraid to work with unfamiliar objects. Always try using new strategies and ways to do stuff—just be safe when blowing things up! Play, eat cookies, have fun, and always remember to not give up when you make a mistake.

FRUITY DNA DIFFICULTY LEVEL

Collect, isolate, and observe DNA from a strawberry.

MATERIALS

Safety glasses
Isopropyl alcohol (rubbing alcohol)
1 large strawberry with leaves trimmed off
Small zipper-sealed plastic bag
Water
Liquid dish soap
Salt
Measuring cups and spoons
Small, clear cups
Spoon
Piece of cheesecloth or coffee filter
Rubber band (optional)
Pipette or dropper (optional)
1 craft stick or toothpick

METHOD

1. Place the isopropyl alcohol in the freezer at least 1 hour before starting the activity and leave it in until step 7.

2. Put on your safety glasses. When working with chemicals, it is always important to protect your eyes!

3. Place the strawberry inside the plastic bag and seal the bag closed. Use your fingers to mash up the strawberry really well for about 2 minutes.

STEP 3

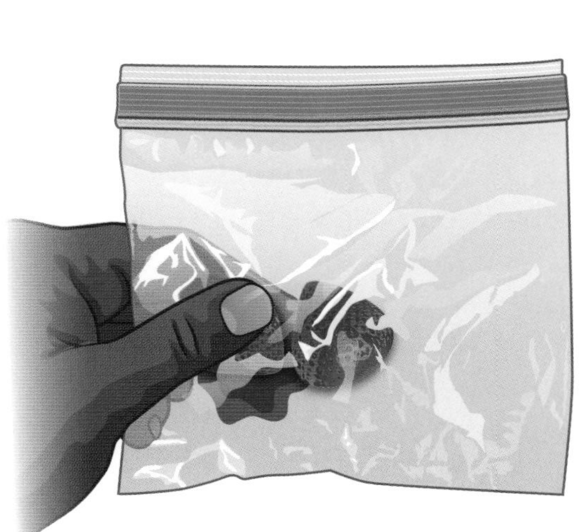

Mash

4. To get the DNA out of the strawberry, you need to break apart the cells that hold the DNA inside. This process is called *cell lysis*. To make the lysis solution that will break apart the cells in the strawberry and release their contents, combine the following materials in a cup, and stir well with a spoon:

6 tablespoons water
1 tablespoon liquid dish soap
1/2 teaspoon salt

5. Pour 3 tablespoons of lysis solution into the bag containing the mashed strawberry. Seal the bag and use your fingers to mix everything together for 1 minute. Let the mixture sit for another minute.

6. Next, you need to filter out the solid parts of the strawberry, leaving just the liquid, which now contains the DNA. Hold the cheesecloth or coffee filter over a cup. Carefully pour the strawberry mixture into the cheesecloth or coffee filter, and gently squeeze the liquid into the cup. It may help to have an extra set of hands for this step. Or you can secure the cheesecloth or filter to the cup with a rubber band. It's not necessary to get all the liquid out, so don't squeeze too hard.

STEP 6

7. Get the cold isopropyl alcohol out of the freezer. Slowly add 1/4 cup isopropyl alcohol to the solution by gently pouring it down the side of the cup. You can also use a pipette or dropper for this if you have one. DO NOT MIX IT! The isopropyl alcohol will sit on top of the mixture of strawberry and lysis solution.

8. Allow the solution to sit undisturbed for at least 10 minutes. Observe any changes you see and record them on your experiment test sheet.

9. After a few minutes, you should start to see a cloudy, white substance forming in the alcohol layer in your cup (the clear top layer). This is the chromosomal DNA extracted from inside the cells of the strawberry.

10. After 10 minutes, you can use a craft stick or toothpick to gently remove the extracted DNA from the alcohol solution.

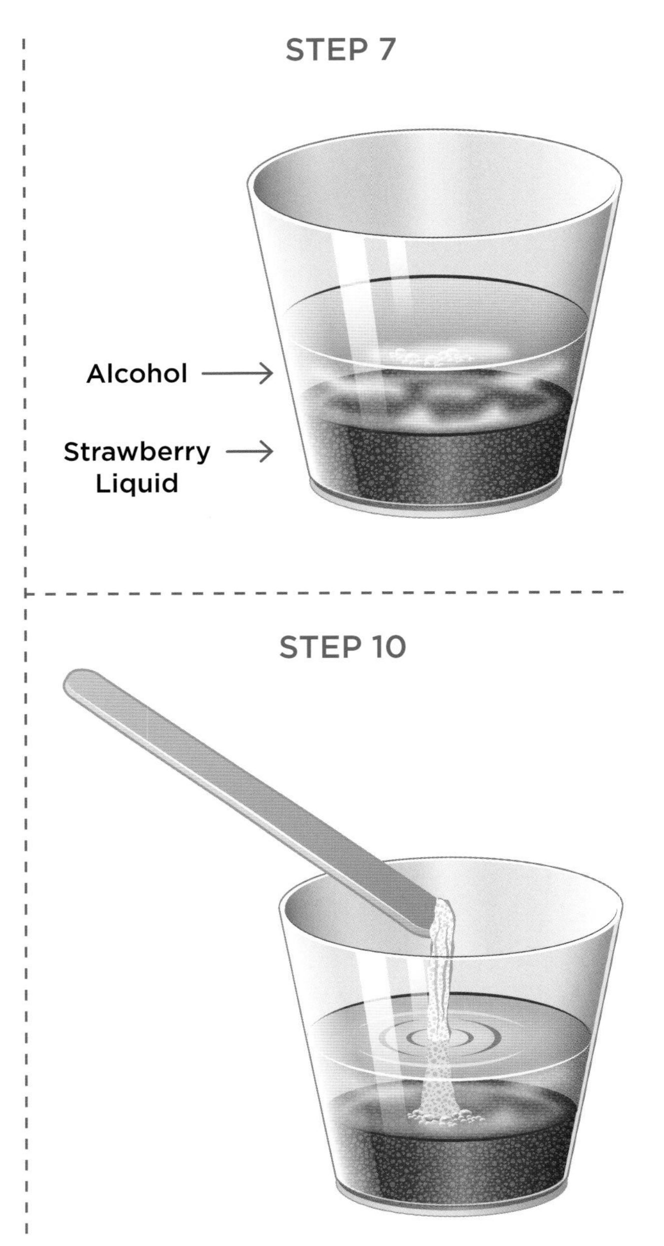

STEM APPLICATION

DNA is inside all living cells and acts like an instruction manual or user code that controls how cells work. DNA stands for *deoxyribonucleic acid*. Every cell—in plants, animals, and bacteria—contains DNA. DNA code is made up of four chemicals, which are abbreviated with the letters A, T, C, and G. They repeat over six billion times in each cell. The complete set of DNA that makes you (and each of us) unique is called your *genome sequence*. Ninety-nine percent of our genome is the same as everyone else's, but it's the other 1 percent that makes us each unique. That 1 percent determines things like your eye color, whether you have curly or straight hair, and which is your dominant hand.

Sometimes there can be glitches in the DNA code, either from birth or as your cells replicate. These glitches, or *mutations*, are often harmless or can be beneficial. Scientists have found that our genome has mutated over thousands of years as we have evolved. Sometimes these glitches can cause health problems. Cancer or muscular diseases are often caused by mutations in our DNA. This is why it is so important for scientists to isolate and understand our genome sequence. This knowledge could lead to a better understanding of what causes certain diseases and help scientists come up with solutions.

If you took all the DNA in your body and stretched it out into a line, it would reach to the moon and back six thousand times—and the moon is 235,000 miles away! To fit all this DNA code into our bodies, it is coiled up into what's called a *double helix* structure. It looks like a tightly packed spring, which allows it to fit inside our cells.

For this experiment, you extracted the DNA from a strawberry. In order to get the DNA out of the cells, you first needed to break apart the cell walls, allowing the contents of the cell to be released. The cell lysis solution you made contained soap and salt. The soap interacts with the strawberry's cell wall and breaks it open, while the salt helps to remove other things like proteins that are attached to the DNA. To get the DNA out of the solution so you can actually see it, you used isopropyl alcohol. DNA does not dissolve in alcohol, which is why it comes out of the solution and clumps together. Scientists can take this DNA and further purify it to analyze and study.

Ellie, age 9

What are some things you want us to know about you?
I have a cat named Pancake, and I like Harry Potter. I also like playing soccer and drawing. My favorite things to draw are animals and trees.

Where do you do stem?
I do STEM activities at home and sometimes at Girl Scouts or summer camps. I have also done STEM activities at school STEM nights and at STEM Like a Girl events.

When did you first become interested in STEM?
When I was trying to get something from under the couch that I couldn't reach, I made an invention to grab it. I used three pencils, an eraser, tape, and a small blue light from an old Valentine gift. I liked inventing something new, and I wanted to do more things like this.

Why do you like doing STEM activities?
I like doing STEM activities because they're fun and interesting. You never know what exactly is going to happen, and I like mysteries.

What is your favorite STEM project you've ever done?
I upgraded my tool for under the couch by adding a small light and a pulley system. It worked just like I thought it would, and I could grab small objects with it.

When an experiment or design doesn't turn out how you expect, how does that make you feel?
I feel okay, because then you just learn what not to do.

What other activities make you feel courageous, confident, and bold?
Playing sports makes me feel confident, like when I dribble the ball down the soccer field and score. I will try almost any arts and crafts, and I like to figure out how to make new things all on my own. I feel courageous when playing one of my favorite games, Catan Junior. Trying to make the right choices or moves to win makes me feel strong.

Who do you look up to?
My mom. She cheers me on when I feel like I can't do something or don't want to.

What are some ways girls can do STEM activities at home?
You can collect items around your house and tinker with them and create something new.

What advice do you have for other girls?
Stick with it and have fun. It is okay to make mistakes because you learn from them.

Maris, age 12

What are some things you want us to know about you?
I go to a public school in my city that's a math and science magnet school. In my free time, I like to draw and bake. I love to play sports, mostly soccer and tennis.

Where do you do STEM?
I do STEM at school and at Girls Count camps.

When did you first become interested in STEM?
I have liked to do mathematics since I moved and went into a new elementary school. I think the teachers made me realize my skill by challenging me to try harder problems. Throughout the rest of my time at that school, I continued to be put in more challenging math groups, which of course helped me want to keep doing it.

Why do you like doing STEM activities?
They usually challenge me to think harder, and I get satisfaction from solving problems.

What is your favorite STEM project you've ever done?
My favorite STEM project I have ever done was for my sixth-grade school science fair. My board was about how the amount of water in a wineglass affects the sound it makes when you rub a wet finger around it (which makes it "sing"). Though I didn't do some of the requirements for winning, I worked very hard on it for a long time and thought it turned out well! I thought it was pretty fun.

When an experiment or design doesn't turn out how you expected, how does that make you feel?
If there are others that succeeded in doing those experiments, I start feeling pretty down because I am very competitive when it comes to most things. But when I'm alone I usually just go to one of my parents and ask them for help.

What other activities make you feel courageous, confident, and bold?
I feel courageous when I'm jumping off of nine-foot ledges into water.

What are some ways girls can do STEM activities at home?
I think online camps like Girls Count are a great way to do STEM activities at home. Also, looking on Pinterest is a great way to find new projects to try.

Who do you look up to?
My teachers have been important mentors and kept me going through tough times. Sometimes you need to wait for the right person to instruct you and understand your skill to help you improve.

What advice do you have for other girls?
Don't be afraid to challenge yourself to try hard things. It's the best way to learn new things.

Julia, age 9

What are some things you want us to know about you?
I have been doing taekwondo for five years, and I am a black belt. I love neon colors, and I love animals. I also really love jewelry and collect different necklaces and bracelets when we go on trips.

Where do you do STEM?
I do STEM at school in science class.

When did you first become interested in STEM?
I became interested in STEM when I first went to a STEM Like a Girl workshop. Everything there was so fun and interesting. I really liked making an art sculpture that moved in the wind.

Why do you like doing STEM activities?
I like them because you can create stuff and learn really cool things. And there are fun projects to try. It was cool to see how being artistic could be combined with building something.

What is your favorite STEM project you've ever done?
My favorite STEM project was making propellers out of sticks, tape, and paper at school. We were testing to see if they spun really fast or did not work. It was fun to test them out and see everyone else's.

When an experiment or design doesn't turn out how you expected, how does that make you feel?
It makes me feel disappointed, but sometimes happy because I can create and add more to it.

What other activities make you feel courageous, confident, and bold?
Two things that make me feel confident are swimming and taekwondo. I like diving and I love the water. In taekwondo, I felt so happy when I learned I passed my black belt exam. Someday I want to be an instructor.

Who do you look up to?
My parents. They help me and encourage me and make me feel happy.

What are some ways girls can do STEM activities at home?
They can use materials they have at home and create stuff. They can also study about things that interest them, like animals.

What advice do you have for other girls?
Try your best, and NEVER give up!

CREATIVE CODING

DIFFICULTY LEVEL

Make a beaded bracelet or necklace using the principles of computer coding.

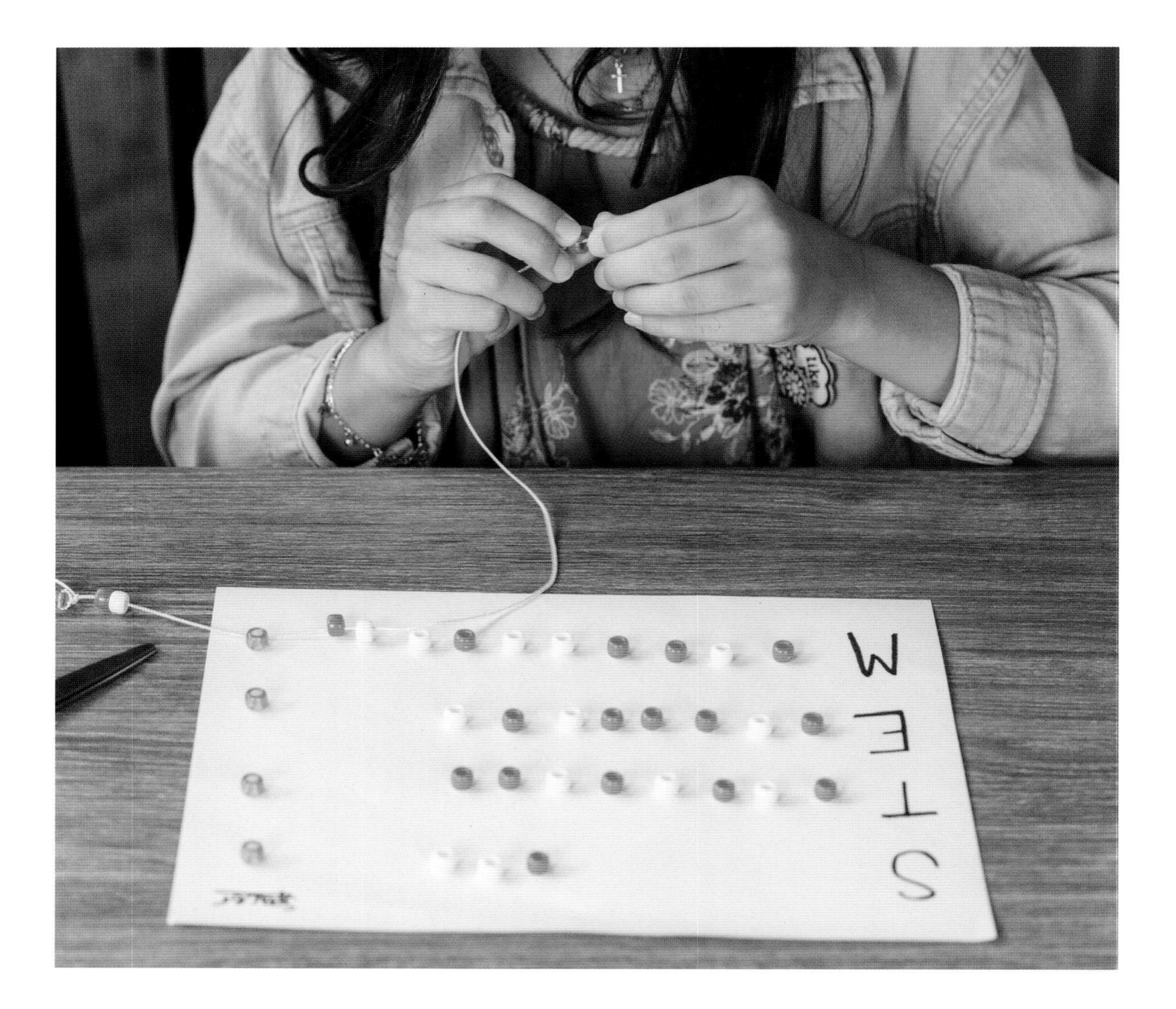

MATERIALS

A handful of beads in three different colors: one dark color, one light color, one neutral color
Elastic thread (make sure it fits through the holes in your beads)
Scissors
White paper
1 marker or pencil

METHOD

1. Cut a piece of elastic approximately 12 to 18 in. long. Tip: It's better to make your elastic too long as it can be trimmed when you are done.

2. Tie one of your neutral-colored beads to the end of your elastic so that the rest of the beads don't slip off when you are making your necklace or bracelet.

STEP 2

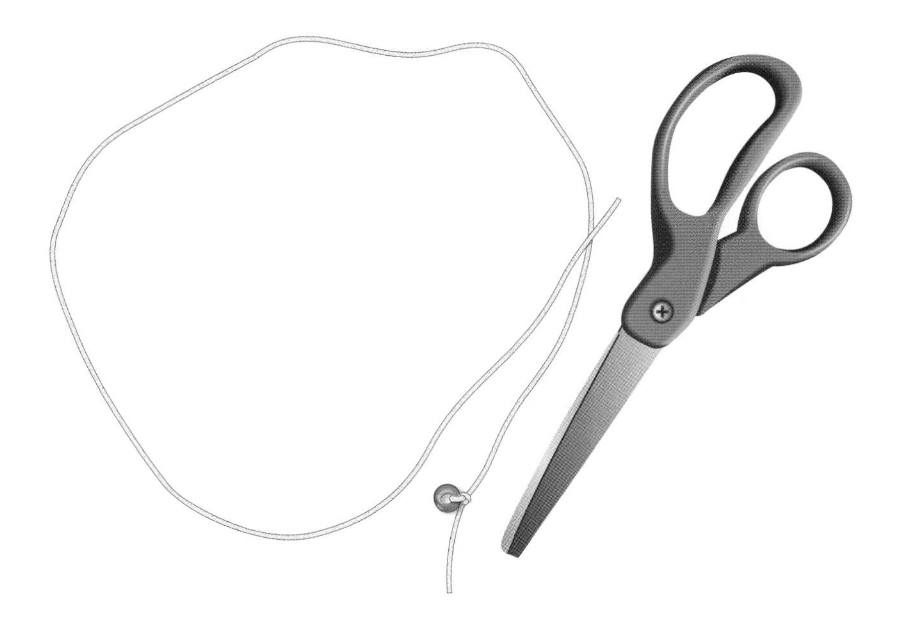

3. Now it's time to write your code. The chart below shows how each letter of the alphabet can be represented by a series of eight light-colored or dark-colored boxes.

You are going to use different-colored beads to spell out the word "STEM" using the following code:

The dark-colored beads will represent the dark boxes for each letter.
The light-colored beads will represent the light boxes for each letter.
The neutral-colored beads will represent spaces between the letters.

STEP 3

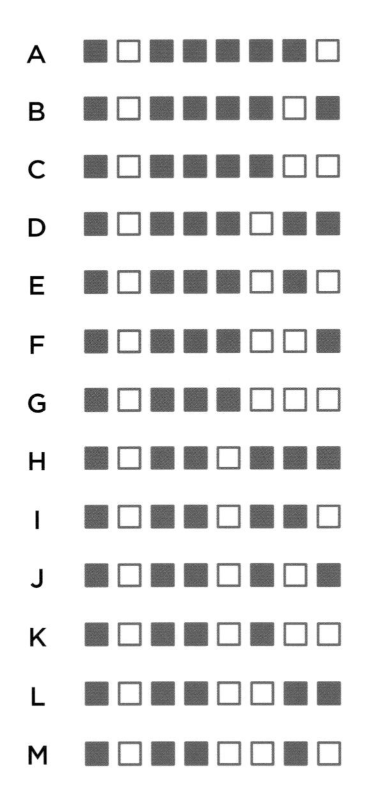

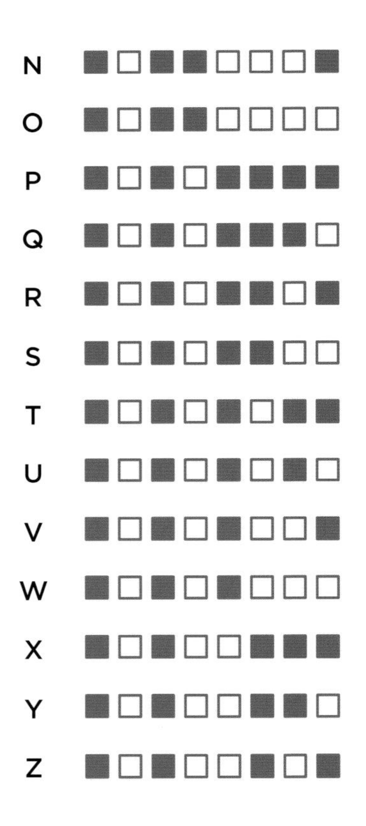

4. Write the letters "S T E M" vertically down the side of your paper. To code the S, find S on the chart. Follow the dark and light pattern with your beads and lay them out on the paper next to the letter S, from left to right. Place one neutral-colored bead at the end of the letter to serve as a spacer.

5. Continue to lay out the beads on your paper with the codes for the letters T, E, and M.

6. When you're finished, your beads should be arranged as follows:

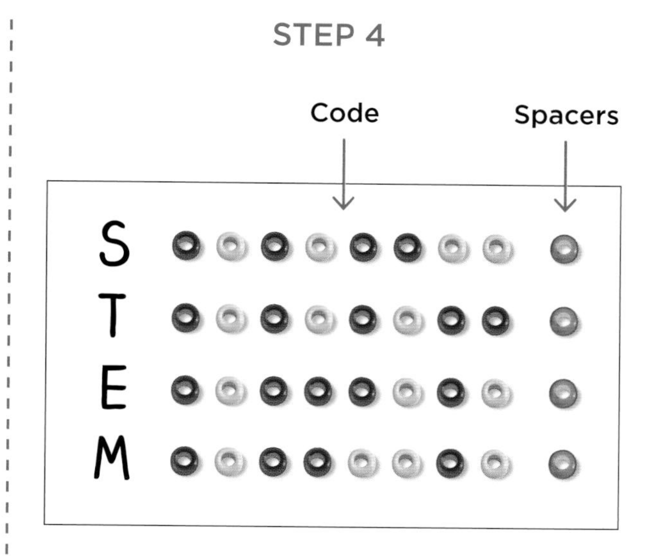

S = dark / light / dark / light / dark / dark / light / light

T = dark / light / dark / light / dark / light / dark / dark

E = dark / light / dark / dark / dark / light / dark / light

M = dark / light / dark / dark / light / light / dark / light

7. Once your code is laid out, start making your bracelet or necklace by stringing the eight beads for the letter S onto your elastic. Begin with the bead closest to the "S" on your paper and work from left to right until you've strung all the beads in order. It's a good idea to double-check your code after each letter, because a small mistake completely changes the message!

8. Once you have all the beads on the elastic to code for S, add one of your neutral-colored beads as a spacer.

9. Continue with the beads for the remaining letters, working from left to right, being sure to add a spacer bead after you finish the code for each letter.

10. When you're done, tie the ends of your elastic together, and wear your coded jewelry. Or you could attach your beads to a zipper on your bag for a coded zipper pull.

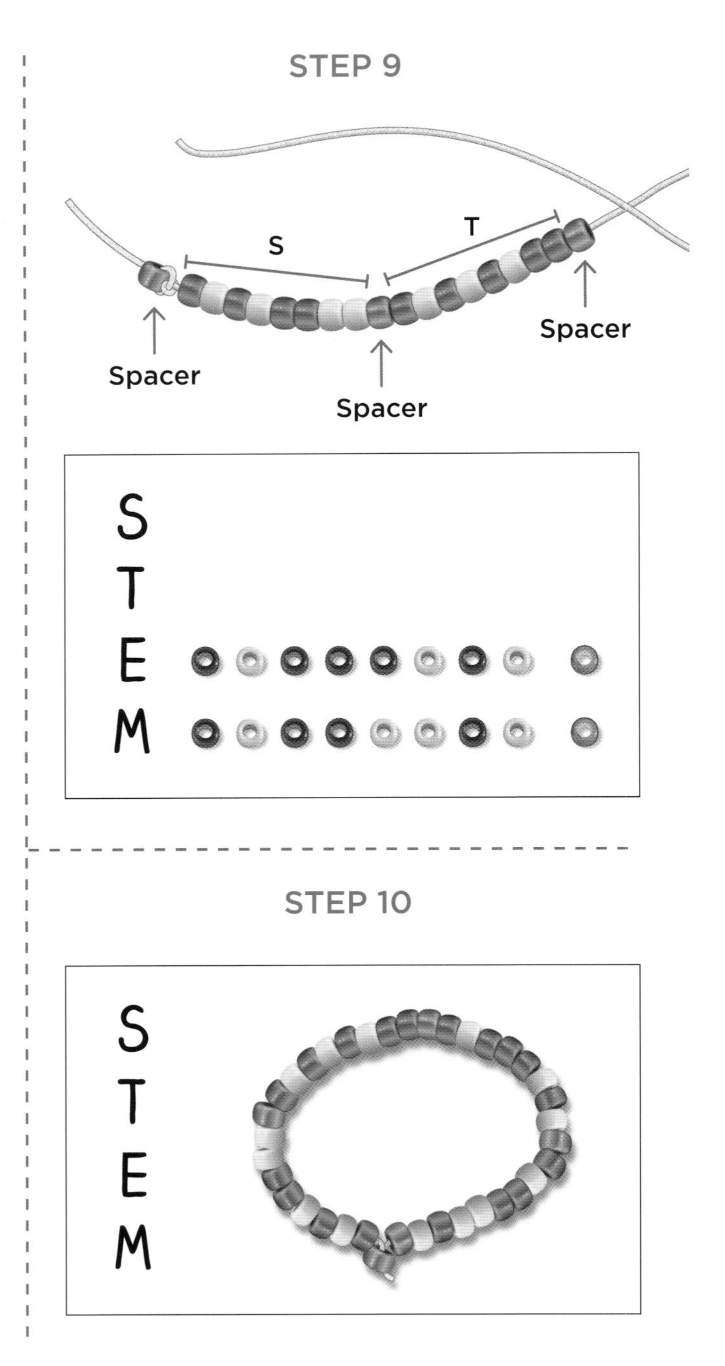

11. Get even more creative and use the code to spell other words—like your name, your initials, "BFF," or another fun word.

12. You can give coded jewelry and a picture of the alphabet code chart to a family member or friend to see if they can decode your message!

STEM APPLICATION

Binary code is a method of representing letters and words using only two symbols ("bi-" means two). Each of the two digits is called a *bit*, which is short for "binary digit." Typically for computer programming, these symbols are zero and one, but there are other forms of binary code you might already be familiar with. Braille is a type of binary code that uses raised or not-raised dots as its two symbols. Morse code uses a series of beeping sounds, written as dashes (long beeps) and dots (short beeps) as its two symbols.

In ASCII (American Standard Code for Information Interchange) binary code, each letter of the alphabet is assigned a string of eight zeros and ones as its bits. This is referred to as an 8-bit code. You can write any word or computer command by "spelling" out the letters with zeros and ones. This type of binary code is used for writing computer-processor instructions, encoding data, and encrypting sound and images onto CDs and DVDs. It can also be used to write secret messages for yourself or your friends to decode! In this activity, the light-colored beads represent the zeros, and the dark-colored beads represent the ones. The neutral beads separate the code for each letter. In computer science, these separators are called *delimiters*.

DESIGN CHALLENGES: INTRODUCTION

The following four activities are called design challenges. They are open-ended projects, which means there is more than one way to solve the problem. You will notice that there are fewer instructions and pictures showing you exactly what to do, so get creative and think like a scientist or engineer to design your own solution. Remember, there is no right or wrong way to complete these challenges. You might do it one way, and a friend or family member might do something totally different. That's okay!

Think back to the design process from the book's introduction. It will be a valuable tool to use as you work through these activities. It's tempting to jump in and start building or experimenting right away, but following the design process helps you think through your work using a well-planned method. Start by asking yourself what problem you are trying to solve. Being clear about what you are working toward will help guide you to design the right thing.

Next, take time to imagine and brainstorm different ideas. Remember that your first idea might not end up being the best, so it's important to come up with a few options. Once you decide on a design to try, it's time to plan it out and select which supplies and materials will work best. As you start creating and building your design, you will test it along the way and make improvements based on what you learn and observe. This is the same process people working in STEM use all the time when faced with an open-ended challenge.

Before you start, remember that it's okay to fail! Not only is it okay, but it's pretty much expected. You might get frustrated if something doesn't work or go the way you thought it would. Failure is a valuable part of STEM, and you can learn so much from those failures. Study what parts of your design aren't working and make observations. Use those observations to improve your design and keep trying. Success will feel even better after you've worked through the tough times!

LET'S RACE DIFFICULTY LEVEL

DESIGN CHALLENGE

Build a car to race across the room using only a balloon to power it.

MATERIALS

1 "car body," such as a large plastic cup, cardboard tube, small cardboard box, empty juice box, empty water bottle
3 straws
2 wooden sticks, such as bamboo skewers or chopsticks
4 "wheels," such as foam balls, wooden spools, Life Saver candies
1 balloon
Masking tape
Rubber bands
Scissors
Craft putty (optional)

IN ADDITION, YOU'LL NEED:
Paper and pencil (to sketch design ideas)

METHOD

1. This design challenge is open ended, so you get to decide how to solve it! Remember to use the design process to think like a scientist or engineer.

EXAMPLES

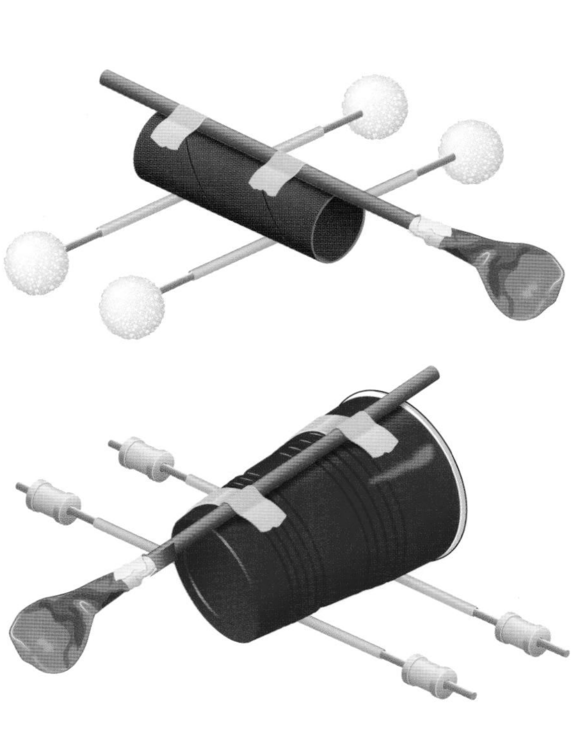

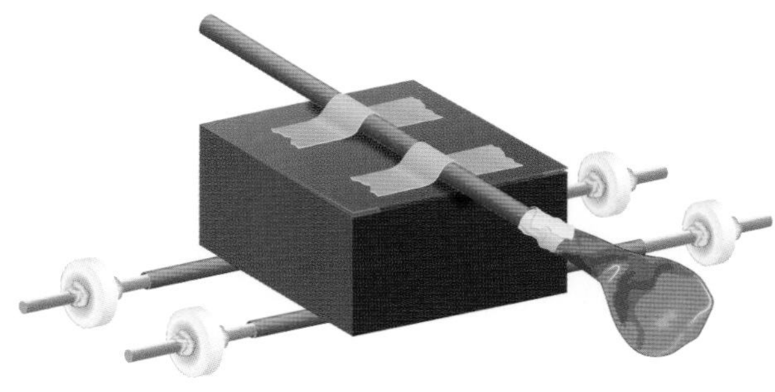

2. Think about the requirements for your car. Although cars can look very different, they have two design features in common: a body for people to ride in, and wheels to make the vehicle roll. Some factors you should consider include:

Weight of the car
Shape of the car body
Size of the wheels

3. Select a body to use for your car. Some ideas are listed in the materials section, but get creative with other things you may have around your house.

4. Axles are what connect the car body to the wheels. Tape one straw to the front of the car body and one to the rear. Insert one wooden stick into each straw. You want part of the wooden stick coming out each end of the straw, so you may need to cut the ends off the straws to make them shorter. Test the axle by rotating the wooden stick inside the straw. Make sure it turns smoothly.

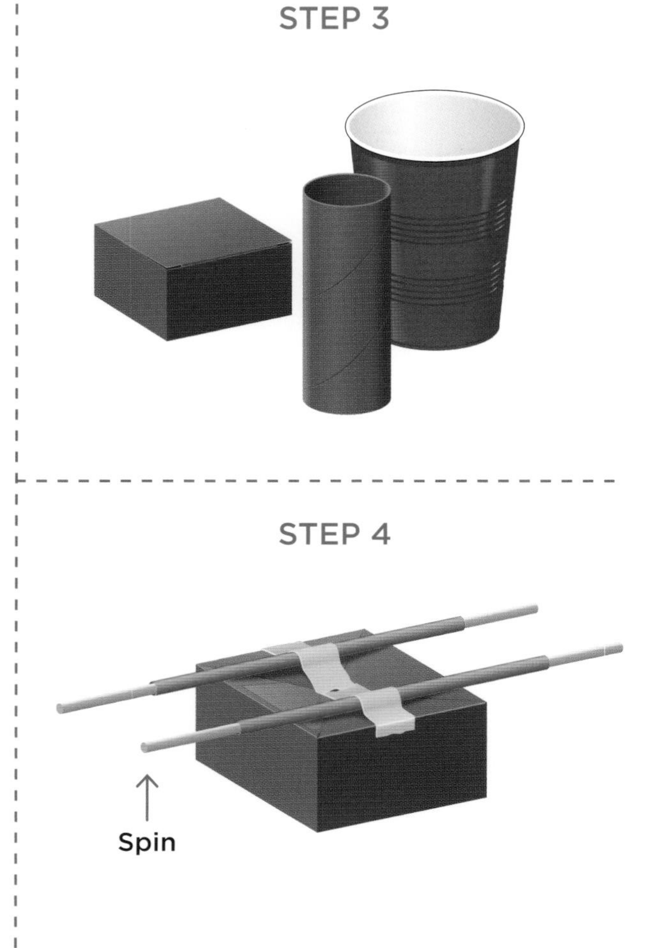

5. Select four wheels. Some ideas are listed in the materials section, but feel free to try out other round objects. Secure the wheels to the end of each wooden stick. Depending on what you use for wheels, you may need to use a small piece of craft putty to secure them to the wooden stick.

6. Give your car a push, and then observe how it rolls. Do the axles and wheels rotate smoothly, allowing the car to move? Make adjustments to your design if necessary.

7. Now it's time to make the power source for the car. Blow up a balloon, let it go, and observe what happens. The air inside the balloon wants to escape, causing the balloon to fly all over the room. In order to make your car go straight, you need the force created when air leaves the balloon to push in only one direction. To do this, tape a straw into the opening of the balloon. Now when the balloon is blown up, the air will be directed out of the straw, making the balloon fly straighter.

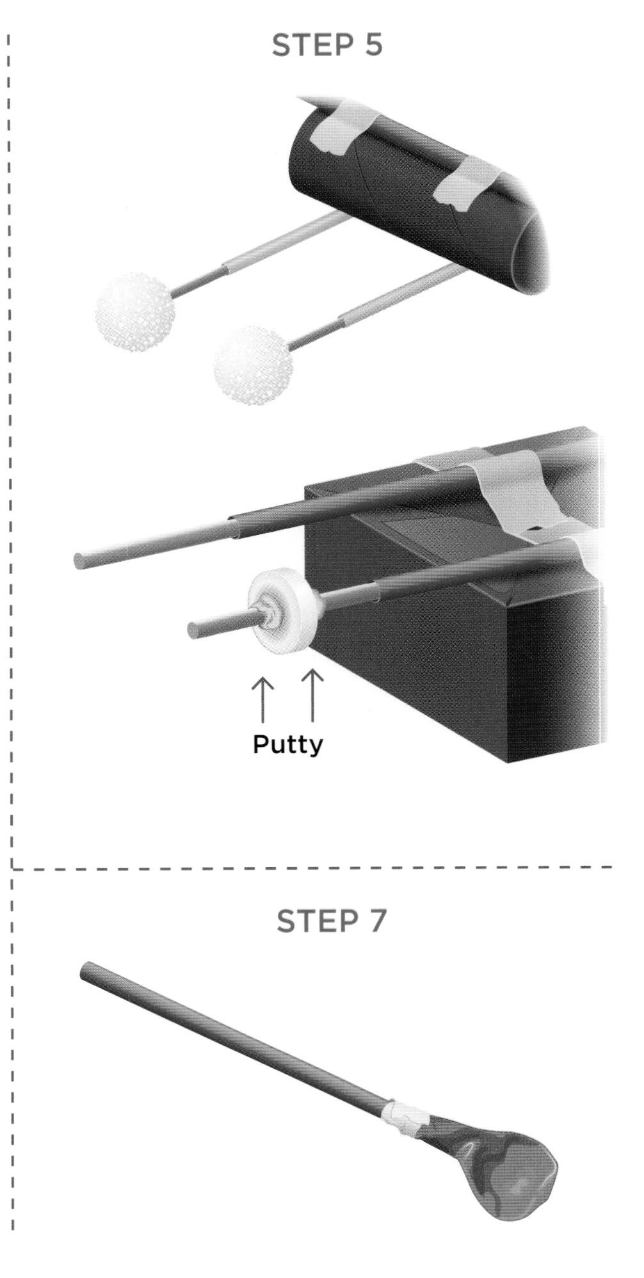

8. Tape the balloon and straw to the top of the car body with the straw's opening pointed toward the rear of the car.

9. Use the straw to blow up the balloon, and then plug it with your finger. Place the car on the ground and release your finger. Observe what happens as the air leaves the balloon. How far did your car travel? Did it travel straight or zoom all over the place? What can you learn from this first attempt? Record your observations on your experiment test sheet.

10. Reevaluate your design. What worked well and what didn't? Make changes to your car, wheels, or balloon that may help it work better.

11. Continue testing and evaluating. Each time, think about what you learned.

STEP 8

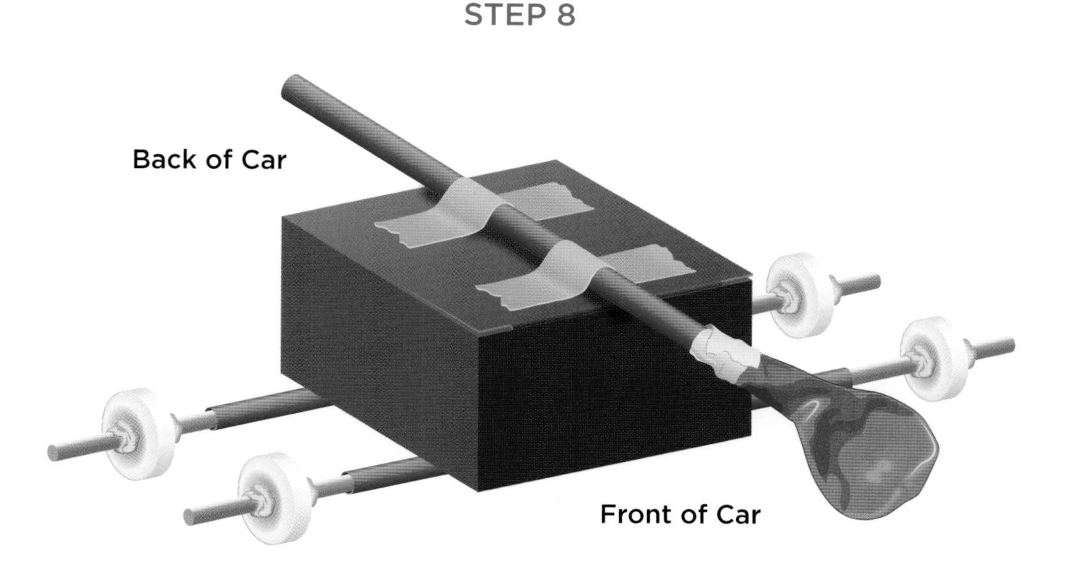

STEM APPLICATION

There are several mechanical forces at work that determine how your car travels along the floor.

Thrust is the force that pushes something forward and keeps it in motion. To create thrust and make your car roll, you need some form of energy such as an engine or motor. In this activity, the energy source is the air pushing out of the straw, creating enough force to make the wheels turn and make your car roll. *Drag* is the force that works in opposition to thrust. As your car moves along the floor and through the air, it creates *friction* (another word for "drag"), which acts to slow down the car. The shape and weight of your car are two things that will affect how big this drag force will be. If you want to make your car start rolling and keep it rolling, the thrust has to be big enough to overcome the drag on your car.

Thrust is based on Newton's third law of motion, which states that for every action there is an equal but opposite reaction. Air squeezing out of the balloon through the straw is the action. The balloon is attached to the car with the straw opening toward the rear of the car, so the escaping air pushes in that direction. Remember that the reaction will be in the opposite direction, so the result is that the car is pushed forward.

How far your car travels is also determined by Newton's third law of motion. Not only is the reaction opposite to the action; it must also be equal in size to the action. The greater the force escaping from the balloon, the more power your car will have. You can experiment with this by changing how much air you blow into the balloon. The more air, the greater the force, which leads to more power to help your car travel farther and faster.

Lena, age 12

What are some things you want us to know about you?
I love to draw, paint, and sketch. I also play basketball for my school team and take tennis lessons. I want to be an engineer or an architect when I am older, after college.

Where do you do STEM?
I do STEM in school and at summer camps. I have a class called iSTEAM where we do experiments. I also do experiments at home with my dad that we find online.

When did you first become interested in STEM?
I became interested in STEM about two years ago when my fifth-grade math/STEM teacher introduced it to me. She always made it fun and we would do group activities and experiments. We got to use the 3D printer, which I really liked.

Why do you like doing STEM activities?
I like doing STEM activities because the engineering of how things work amazes me. I like to figure out how things work.

What is your favorite STEM project you've ever done?
My favorite STEM project I've ever done was designing a tiny house. I used a computer program to design it and to figure out measurements. Then I did some math and figured out some more measurements, so the measurements on the computer could be converted to the sizes of random objects I found to build it with.

When an experiment or design doesn't turn out how you expected, how does that make you feel?
It is frustrating at first, but then it just gives me another opportunity to fix it and learn from it.

What other activities make you feel courageous, confident, and bold?
Drawing makes me feel confident. I love when the picture comes out just right or when the new skill I'm learning, like shading, comes out right. I'm also getting better at playing basketball and working with a coach, so that is making me feel more confident. Helping others also makes me feel good.

Who do you look up to?
I look up to my dad. He is an engineer and he is so creative in everything, even simple stuff like fixing something around the house. Everything he does, he does it in a unique way.

What are some ways girls can do STEM activities at home?
Some ways girls can do STEM activities at home are by looking up videos online or just looking around their house for something to be creative with.

What advice do you have for other girls?
Be creative, work hard, have fun, and don't let failure stop you!

HELPING HANDS DIFFICULTY LEVEL

DESIGN CHALLENGE

1. Build a device that can pick up a cup from one foot away and put it back down without damaging the cup.

2. Only one person can operate the device at a time.

MATERIALS

Ideas for materials to use for your device are listed below. You don't have to use them all, and you can get creative with other things you have around the house.

Craft sticks
Clothes pin
Rubber bands
Paper clips
Straws
Pipe cleaners
Plastic spoons

IN ADDITION, YOU'LL NEED:

Paper and pencil (to sketch design ideas)
Ruler
Tape
Scissors
Small paper or plastic cup
Several small objects, such as coins, a pencil, cotton balls

METHOD

1. This design challenge is open ended, so you get to decide how to solve it! Remember to use the design process to think like a scientist or engineer.

EXAMPLES

2. Think about the requirements for your helping hands. What does it have to accomplish?

3. Before you start building, look at your materials and imagine different ways they could be used. Brainstorm a couple of different ideas, because sometimes your first thought doesn't end up being the best.

4. Sketch out your design and decide what materials you'll use to build your device. You don't have to use them all, so think about what supplies would work best based on your design.

5. After you've planned out your initial design, start building. As you build, do you find you have to make changes to your initial plan?

6. Once you have your first device, test it! Use your ruler to measure 1 foot and place the cup at least that far away.

Can your device pick up the cup?
Can it put it back down?
Did the cup get damaged?
What did you learn from this first attempt?

STEP 6

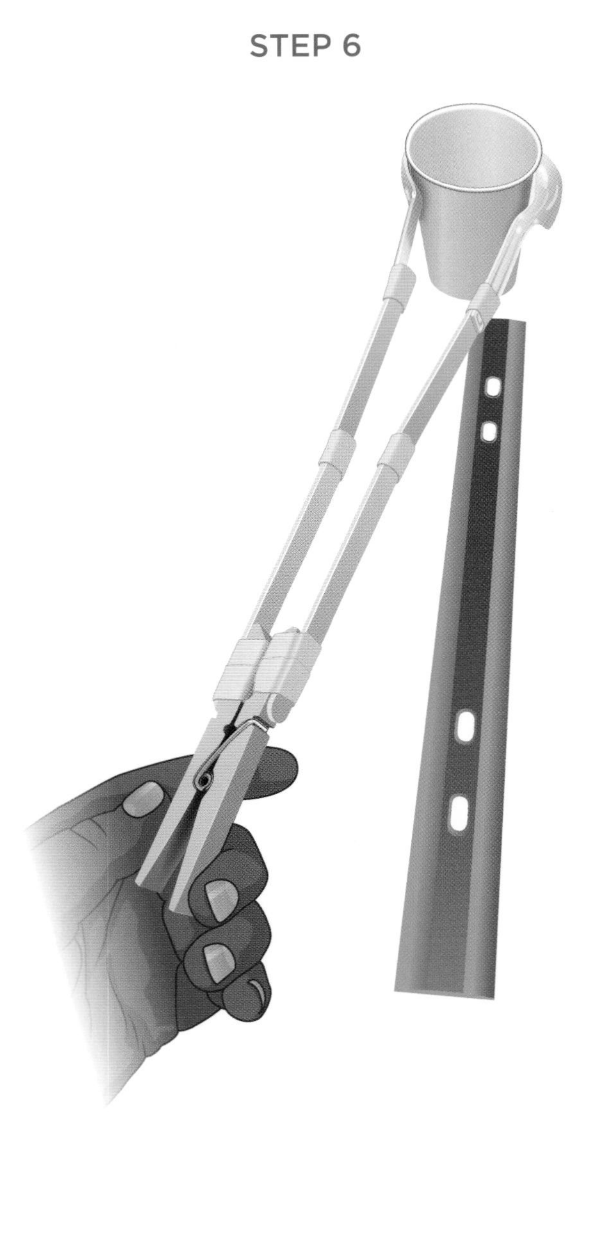

7. Evaluate your design. What worked well and what didn't? Record your observations on your experiment test sheet. Make changes to your device that may help it work better.

8. Continue testing and evaluating. Each time, think about what you learned and how it helps you improve on your design.

9. Once you have successfully completed this challenge, see if you can use your helping hands to complete other tasks (ideas below). Do you have to change your design to perform these additional functions?

- Fill your cup with some small objects like a few coins. Then see if you can successfully dump them out of the cup in a certain spot.
- Can your device pick up other objects, like a pencil, a cotton ball, or ordinary items sitting on a table?

STEP 9

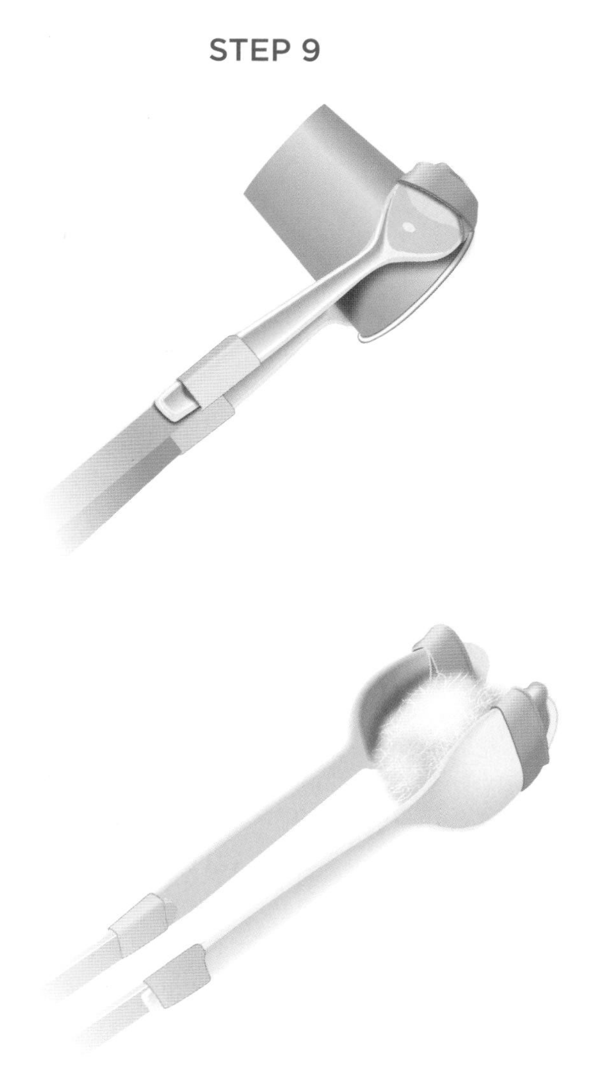

STEM APPLICATION

Imagine having no hands. Make fists with your hands, and try to pick up a pencil, zip your coat, or get yourself a drink of water. Often we take these everyday tasks for granted, but for many people with injuries or disabilities they can be a challenge. *Prosthetic devices* are artificial body parts that replace a missing or damaged part of someone's body. Examples of prosthetic devices that people use include arms, legs, hearts, teeth, and bones. Prosthetic devices can either be implanted into the body, like an artificial heart or a joint replacement, or they can be attached to the outside of the body, like an artificial limb. These devices allow people to live longer and happier lives and to have more independence.

Engineers and scientists often get inspiration from humans or animals. They try to mimic certain features when designing prosthetics. For example, they might study how a cheetah runs if they are trying to design a prosthetic leg for a sprinter. If they are trying to design an artificial joint or bone, they will need to understand the mechanical properties of actual bone to determine what materials would provide the same strength and stability for the patient.

Biomedical engineers, mechanical engineers, and material scientists work together to design prosthetic limbs with a specific task in mind (for example, running, lifting, writing). This is why knowing the exact problem you are trying to solve is so important (the *ask* step in the design process). A prosthetic leg for someone who wants to ride their bike needs to be designed differently than one for someone who wants to walk or run. Did you need to make changes to your design in order to allow your prosthetic arm to dump objects from the cup or to pick up something else, like a pencil?

Devon, age 9

What are some things you want us to know about you?
I like to use chopsticks with my left hand even though I'm a righty. I like to collect small things like gemstones, rocks, and shells. I have dyslexia, which makes it hard for me to do things like read and write. How it works is when you're young, letters move around, but as you get older, you teach your brain to not do that. But it is still hard. Thankfully, I am really good at listening, so I can remember everything I hear when I listen to it. I love listening to books. Sometimes I feel like I am better at things other kids aren't good at, like seeing things in different ways. I think it is called different perspective. I also like to make everything perfect, so I always take my time and do something until it is actually done. I have a really good memory, and I know a lot of things that other kids and parents don't, like I know a lot of weird facts about animals, about things that happened in history, or about tornados and rocks. I am really good at subjects like social studies, math, science, and gym. Some kids in my class complain about these subjects, but I like them!

Where do you do STEM?
I do STEM in my garage, in my backyard, and pretty much all over my house. We were about to finally start a project at school, but then we had to go home for coronavirus. The teacher still sent the projects, and we did them at home.

When did you first become interested in STEM?
Last year we had an assembly where there were all these different STEM projects, and we got to try a lot of them, so that's when I started to like STEM. My favorite was building a huge arch that you had to be able to walk through. I liked building it, and then you got to destroy it when you were done!

Why do you like doing STEM activities?
They're fun. I like to see how things turn out in the end.

What is your favorite STEM project you've ever done?
My favorite STEM project I've ever done is building a sidecar with my grandpa for my dog on my go-kart. I got to use things that I don't usually get to use, like power tools. I even got to use a saw!

When an experiment or design doesn't turn out how you expected, how does that make you feel?
When something doesn't turn out how I expected, I feel like I have to do it again until I get it right.

What other activities make you feel courageous, confident, and bold?
I like doing stuff that lots of kids wouldn't do because they probably think it's not safe, like climbing trees, climbing rocks, hiking, climbing

waterfalls, and swinging on rope swings. I feel strong when I do these things. Like I can do anything!

Who do you look up to?
I look up to my mom because she never stops. If somebody tells her she doesn't need to do something or that nobody else is doing it, she does it anyway even though it's pretty hard. I think she likes the challenge. And she likes to do a lot of stuff that I like to do. I also look up to Richard Branson. He has dyslexia, and now he's making a way to let people who aren't astronauts go on rockets into space. I also look up to Bear Grylls. He isn't scared of anything.

What are some ways girls can do STEM activities at home?
Girls at home can draw a picture of something they want to build and then just do it. Find stuff around the house or yard and use that to build things. Sometimes I collect things that I think I might be able to use later.

What advice do you have for other girls?
You can do whatever you want even if somebody says, "No, that's something that boys do." Some people say you shouldn't do that because you're a girl, but I think you should do it anyway.

WINDY ART DIFFICULTY LEVEL

DESIGN CHALLENGE

1. Combine art and engineering to create a sculpture, called a *kinetic sculpture*, that moves in the wind.

2. Your sculpture needs to stand on its own.

3. Your sculpture should be visually interesting.

MATERIALS

Sculpture base, such as a piece of cardboard or Styrofoam

Ideas for sculpture materials are listed below. You don't have to use them all, and you can get creative with other things you have around the house:

Aluminum pans
Aluminum foil
Ribbon
Small metal washers
Ping-pong balls
Craft foam
Cardboard tubes
Paper cups
Paper clips
Straws
Craft sticks
Wooden skewers
Pipe cleaners
Yarn or string

IN ADDITION, YOU'LL NEED:
Paper and pencil (to sketch design ideas)
Tape
Scissors
Hair dryer (optional)
Hot glue gun (optional)

EXAMPLES

METHOD

1. This design challenge is open ended, so you get to decide how to solve it! Remember to use the design process to think like a scientist or engineer.

2. Kinetic sculptures have at least one feature that moves. For this design, you will be using wind energy to make your sculpture move. Think about the requirements for your sculpture. How will it be supported so it doesn't fall over in the wind? What pieces will move (kinetic), and what will remain still (static)? How will you make your sculpture both functional and interesting to look at?

3. Before you start building, look at your materials and imagine different ways they could be used. What materials would work best as the support for your sculpture to secure it to the base? How could you use the different materials to make them move when blown by the wind? Brainstorm a couple of different ideas because sometimes your first thought isn't the best.

4. Sketch out your design and decide what materials you'll use to build your sculpture. Try to include at least two parts that will move in the wind.

5. Once you've decided on your initial design, start building. As you build, do you find you have to make changes to your initial plan?

SAFETY NOTE

If you are using hard objects like ping-pong balls, a serrated knife or small saw can be helpful. Ask an adult in your house for help first.

6. When your initial sculpture is built, test it by using a hair dryer to simulate wind, and make observations. If you don't have a hair dryer, just use your breath to blow on it. What did you learn from this first attempt? Record your observations on your experiment test sheet.

7. Evaluate your design. What worked well and what didn't? Make changes to your sculpture that may help it work better.

8. Continue testing and evaluating. Each time, think about what you learned and how it helped you improve on your design.

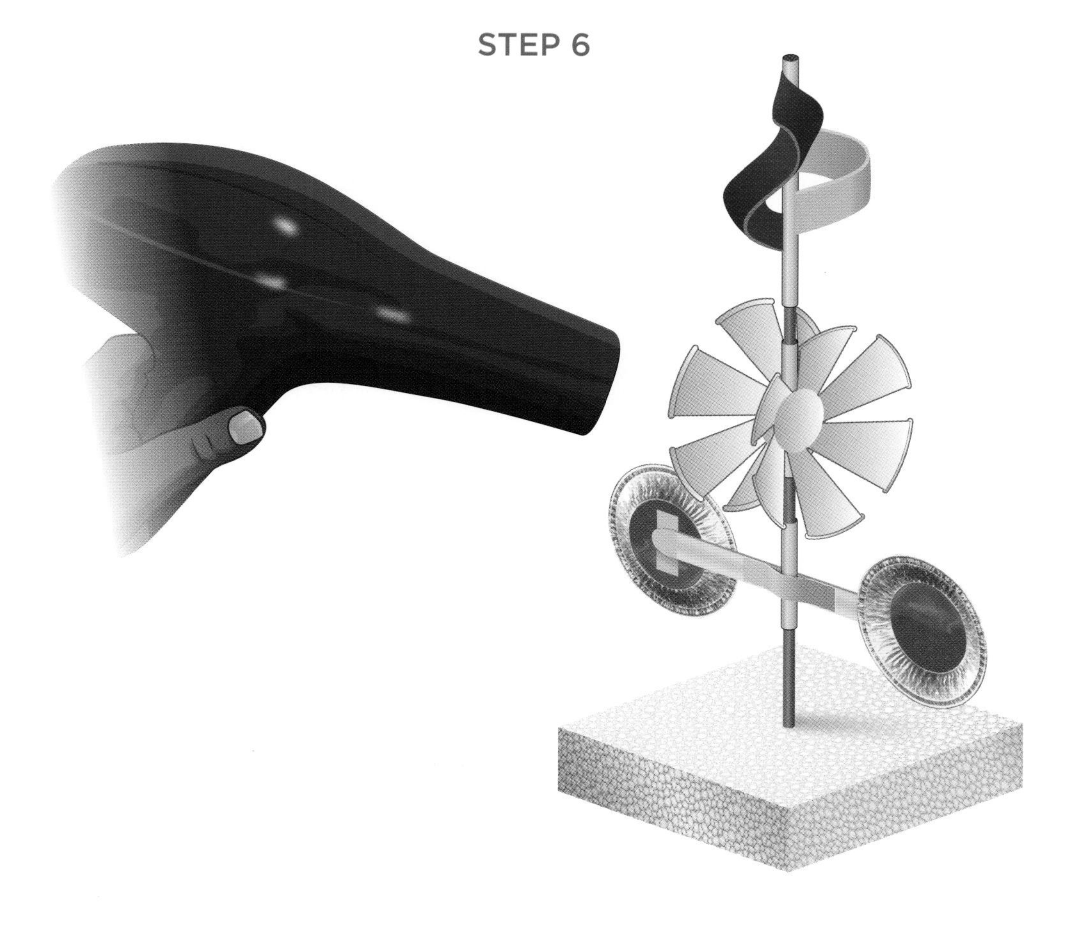

STEM APPLICATION

Can art and engineering go together? A lot of times we think of art and engineering as separate fields: Engineers make useful and practical things, and artists make visually attractive things. But what if we combine the two? Leonardo da Vinci is best known for his paintings and sculptures, but did you know he was also an engineer? He wanted to use art to create new machines, and many people refer to him as the first "systems engineer." He wanted to see how each piece of a machine worked and how those different elements could be combined into a system.

Civil engineering is a broad field that includes building everything you need for a civilization to function smoothly. This would include things like buildings, roads, bridges, and drinking or wastewater systems. Structural engineering is a branch of civil engineering that designs structures. Structural engineers are responsible for designing the skeleton of a building, bridge, or other structure to make sure it is strong and safe.

We think of structures and art as being stationary, meaning they don't move. But some artistic sculptures are designed specifically to move as part of the design. They are called kinetic sculptures. The motion can come from several sources. A motor, a wind-up mechanism, or the wind are all examples of sources to power the movement of a sculpture. As you may have experienced while building your own sculpture, these kinetic sculptures require a lot of engineering and art. The base needs to be strong and stable to support the sculpture in the wind. The pieces that move need to be designed to capture the wind when it blows in any direction. And then to make it artistic, the sculpture needs to be visually interesting for people to look at.

Abby, age 9

What are some things you want us to know about you?
I am in fourth grade, and I am hoping to be part of the salmon project at school. They take salmon eggs and put them in a large tank in the library, where we take care of them. Then, fourth graders get to go on a field trip to the salmon hatchery to let them go. I love to sing and act because those activities make me happy. I love performing. I have a sister named Tessa, and she is in first grade.

Where do you do STEM?
I do STEM at my school's STEAM night every year. At home, I like to make inventions and glue stuff together.

When did you first become interested in STEM?
I started by observing all the different types of pets we have at home. We have three pet snakes named Tummy, Henna, and Lentil. I also have a crested gecko and a tarantula. I used to have a bearded dragon too. I like my snakes because they are nice and used to me. The tarantula is creepy.

Why do you like doing STEM activities?
I like doing STEM activities because I get to be creative and invent. I like going to different places and finding new materials. At my house, I have a secret hideout for inventing, where I keep glue, string, paper, glitter, Popsicle sticks, regular sticks, mirror paper, glass jars, tape, scissors, toy cars, and nail polish. I start a plan in my head of what I am going to do. Then I gather all the materials and finally decide what to make. If I have to change my project because I don't have something, that's okay. I just start working on my project until I'm done. Then I start on a new project.

What is your favorite STEM project you've ever done?
I invented the Weatherio. It is a device that helps you know what the weather is like. I inserted a straw into the middle of a bottle so that when it's windy, you can find out which way the wind is blowing because the straw will move with the wind. There is a clear bucket with measurement marks to see how much rain there is. If it fills up, I know that it is really rainy. Then, there is a cup with foil that reflects the sunlight and tells you how sunny it is and how cloudy it is. I painted all the parts except the foil and covered them in glitter. I then glued it all to a board and wrapped the board in plastic wrap to make it waterproof.

When an experiment or design doesn't turn out how you expected, how does that make you feel?
When my experiment doesn't work out as planned or how I wanted it to go, I get excited. I know that I can just start again with all new materials and try new stuff. Like one time I was trying to make the Weatherio, I tried to attach the straw, bucket, and sun catcher with a glue stick, but it wasn't strong enough. Then I tried Elmer's glue. It failed too. I decided, since neither of them were

strong enough, I needed to use hot glue. It worked!

What other activities make you feel courageous, confident, and bold?
Performing makes me feel happy. I've been singing since I was two.

Who do you look up to?
My mama because she inspires me so much. She is a STEM teacher. She always does activities that inspire me to do DIYs. I like when we make structures out of dry spaghetti and mini marshmallows.

What are some ways girls can do STEM activities at home?
Inventing, DIYing, and STEM crafts. You need lots of materials to do STEM crafts. Materials are 50 percent of STEM activities. The other 50 percent is hard work.

What advice do you have for other girls?
Never give up. Giving up is just another way to say you failed. If you keep trying, you can eventually say that you succeeded.

RAIN, RAIN, GO AWAY! DIFFICULTY LEVEL

DESIGN CHALLENGE

1. Design a model city to keep pollution and water runoff out of the river by filtering out pollution and allowing more water to be absorbed into the ground.

2. Your city needs to have housing for people to live in and roads to connect everything.

MATERIALS

Large, disposable, rimmed aluminum cookie sheet (15 x 10 1/4 x 3/4 in.)
Sponges
Cotton balls
Pieces of cardboard cut into rectangles approximately 2 x 3 in. and 4 x 4 in.
Aluminum foil
Masking tape
Brass brad fasteners
Sprinkles or crumbled cookies
Watering can or large cup
Water

IN ADDITION, YOU'LL NEED:
Paper and pencil (to sketch design ideas)

METHOD

1. This design challenge is open ended, so you get to decide how to solve it! Remember to use the design process to think like a scientist or engineer.

2. The goal is to design two different cities and compare how successfully they slow rainwater runoff and filter pollution, preventing it from washing into the sewers, where it eventually finds its way to streams, rivers, and oceans. (See the "STEM Application" section on page 226 for more information about parks, bioswales, and rain gardens.)

Apartment
House
House
House
House

Design 1
You work for a city that hasn't prioritized adding green spaces to prevent water runoff. Your city (cookie sheet) must have at least four houses (small cardboard pieces), two apartment buildings (large cardboard pieces), and roads connecting all the living spaces (masking tape), but you only have the budget for one park or bioswale (sponge) and two houses built with rain gardens (cotton balls).

Design 2
The city learned that preventing water runoff and pollution is important and has decided to budget for as many green spaces as you need. Your city must still have the same number of houses, apartment buildings, and roads connecting everything, but now you can include as many parks, bioswales, and rain gardens as you want.

STEP 2

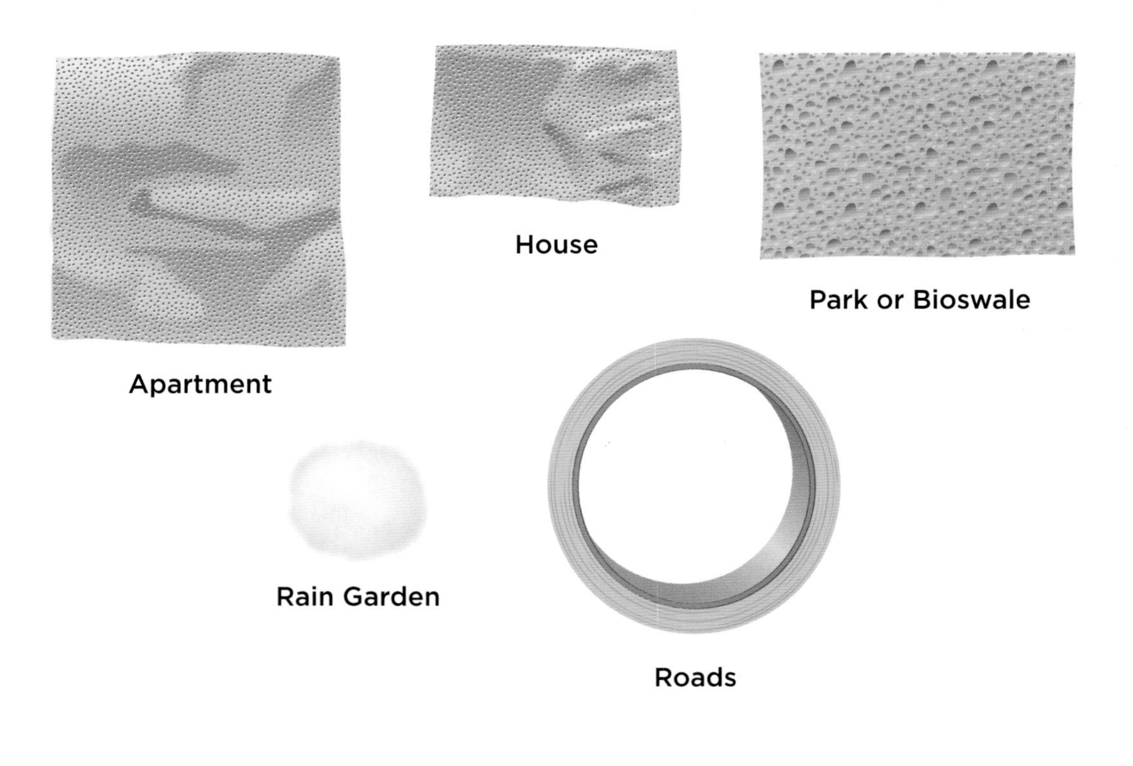

3. Before you start building, sketch out how you plan to lay out your city. Note: The river will flow from the top of the tray.

4. Wrap the pieces of cardboard with aluminum foil so they don't get soggy. Make at least two larger pieces for apartment buildings and four smaller pieces for houses.

5. Start by following Design 1. Use tape to attach the cardboard pieces to the cookie sheet. Use brass brads to attach the sponge and cotton balls. You can use a pencil or a nail to poke a hole through the cookie sheet to poke the brads through. Be careful, as the metal may be sharp.

6. Use the masking tape as roads to make sure everything in your city is connected.

7. When you are ready to test your design, carry your cookie sheet over to the sink. The sprinkles or cookie crumbs represent pollution. Sprinkle some all over your city.

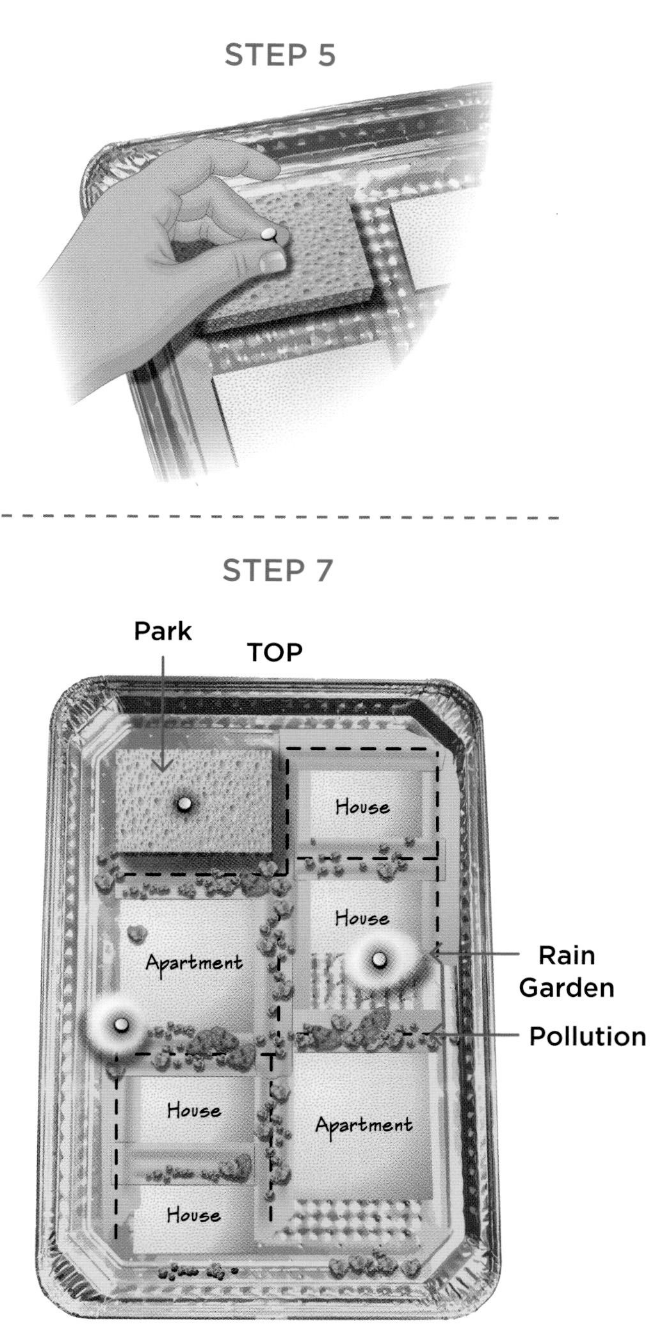

8. Hold the cookie sheet at a slight angle over the sink. Starting at the top of the cookie sheet, use a watering can or large cup to pour water on your city to simulate a rainstorm. It may help to have someone else hold the cookie sheet while you pour the water. Observe what happens to the water runoff and pollution. Did the sprinkles or cookie crumbs wash through the city and into the bottom of the pan (which represents the river), or were they trapped in a park, bioswale, or rain garden? Evaluate what worked and what didn't. Record your observations on your experiment test sheet.

STEP 8

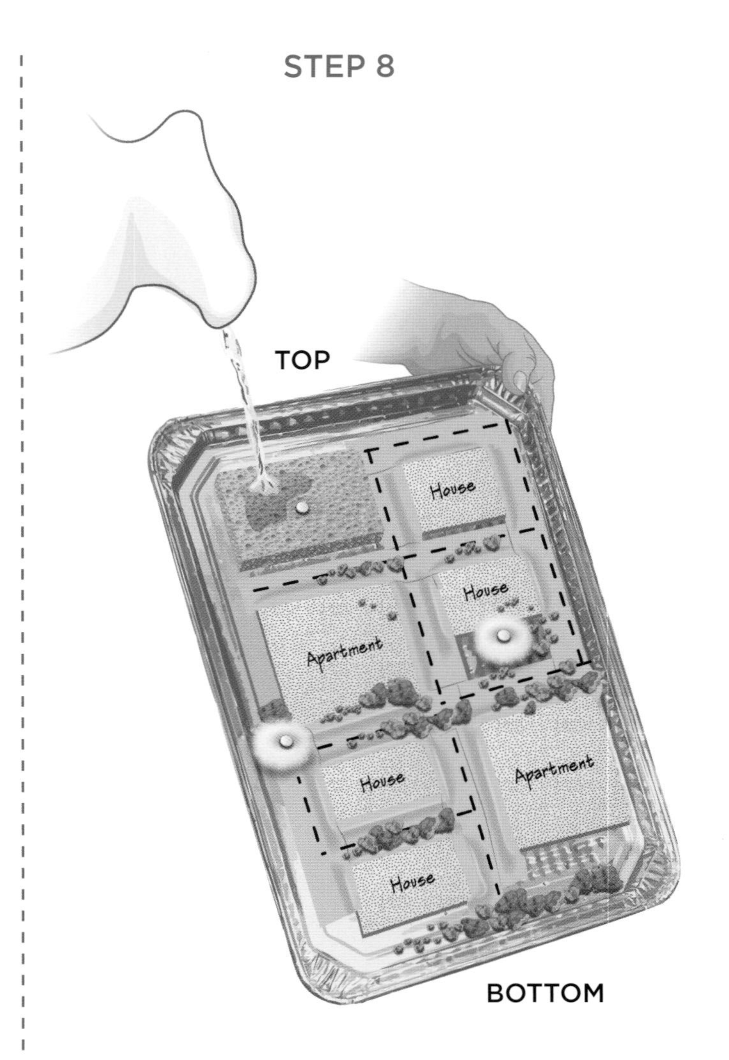

9. Now you are able to redesign your city following Design 2. You can use as many green spaces as you want. Include extra parks, bioswales, and rain gardens.

10. Test your second design the same way, with pollution and a rainstorm. Compare your results to those from your first design. Did the increased number of green spaces help stop the pollution from washing through the city?

STEP 9

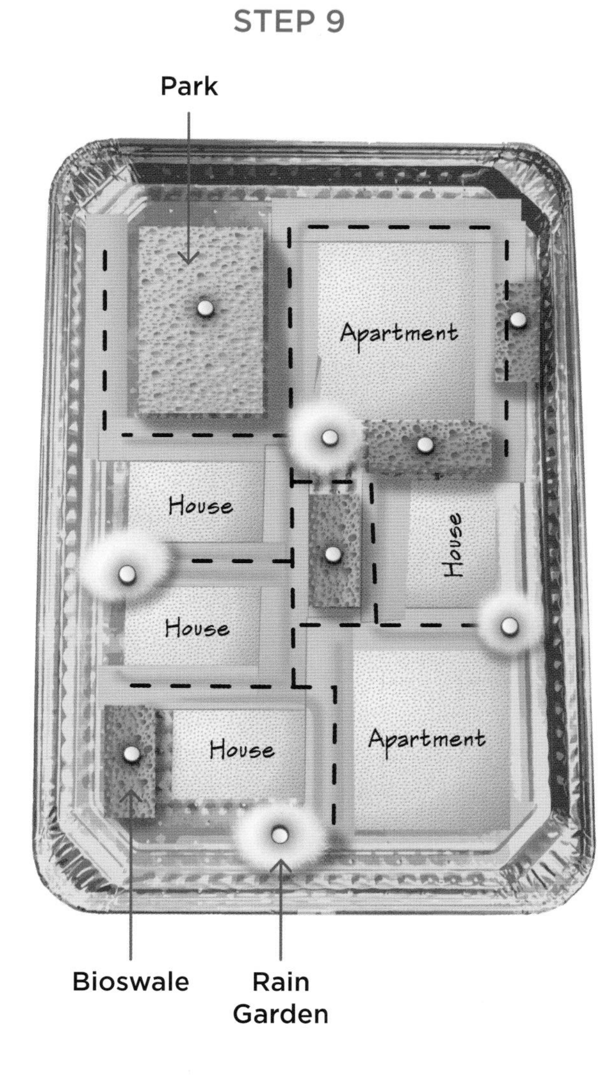

STEM APPLICATION

What happens when there is heavy rain? In a forest or areas with lots of green space, the soil acts like a sponge by absorbing the water. The water can then slowly drain out while the soil acts like a filter, removing pollutants before the water reaches our rivers and streams. By contrast, cities are full of streets, parking lots, and buildings that are made from materials that don't absorb water. When there is heavy rain in a city, the water often flows too quickly over roads and roof tops, which can cause flooding. The water picks up pollutants off these hard surfaces that can end up in our rivers and streams. This is called *urban runoff*.

Water resource engineers design methods to absorb and filter this urban runoff, allowing it to slowly be released into the sewer drains. They can mimic what nature does by including different types of green spaces in city designs. This can be done on both a large and a small scale. *Parks*, which have lots of grass and dirt, act as natural sponges to slow down and filter the urban runoff following heavy rain. *Bioswales* are roadside green spaces that collect water during heavy rain. They look like concrete boxes built lower into the ground and filled with different types of soil and plants. You might have seen bioswales along sidewalks or roads in your town and thought they were there just to look nice. But actually, these bioswales help collect the rain along the roads to filter out pollution and slow down the water flowing to sewer drains. At homes and apartment buildings, people can design *rain gardens* to soak up water runoff from roofs and driveways. Rain gardens are depressed areas in your yard that contain soil, grass, or plants. The water can collect in the rain garden instead of flowing directly down the street, picking up pollution as it goes. Some buildings even have gardens and green spaces on their roofs!

It is important that civil engineers, water resource engineers, and city officials work closely together when planning and designing the layout for a city. Adding green spaces such as parks and bioswales costs money and decreases the number of houses and other structures that can be built. But a lack of green spaces leads to increased urban runoff and polluted rivers and streams, which can harm the ecosystem and cost money to clean up. Individuals can help by planting rain gardens around their houses or apartment buildings that can absorb some of the runoff before it flows into the sewer drains.

Liv, age 8

What are some things you want us to know about you?
I like the colors pink and coral. I love basketball and any other types of sports. I love wrestling with my dad and my sister. I also love to build with Legos. Right now, my sister and I are building a mansion together. My birthday is also on leap day!

Where do you do STEM?
I do STEM in school, at Girl Scouts, and at home.

When did you first become interested in STEM?
I first became interested in STEM when I started putting together things and making stuff. We did a crystal-growing project in second grade at school, and I do STEM projects in Girl Scouts. I like to experiment and mix things together at home in the kitchen and build stuff with my dad.

Why do you like doing STEM activities?
I like doing STEM activities because you can let your imagination go wild! My friend and I like to mix liquids or foods to experiment, like melting marshmallows and chocolate together to make our own s'mores.

What is your favorite STEM project you've ever done?
My favorite STEM project that I've ever done was with my friend. We mixed together bath bombs and other fun stuff. It turned out to be a big mess!

When an experiment or design doesn't turn out how you expected, how does that make you feel?
I feel a little bit disappointed, but I tell myself to keep trying until I get it right! Sometimes it takes a while to come up with ideas, but that's okay. I also work with my dad so that we can come up with ideas together.

What other activities make you feel courageous, confident, and bold?
I feel proud when I score a basket in basketball and everyone cheers and says, "Yeah, Liv!" I also feel confident when I make slime and it turns out good!

Who do you look up to?
I look up to my mom! She finds so many fun STEM projects for me to do. Also, I look up to anyone who has had trouble and fixes it.

What are some ways girls can do STEM activities at home?
Try new things like mixing together foods that don't match, building whatever your imagination leads you to, and making an obstacle course to race with your friend.

What advice do you have for other girls?
If you feel like giving up, then keep trying or do something new, because you might surprise yourself. Happy STEMing!

Natalie, age 10

What are some things you want us to know about you?
I love sports, especially soccer and tennis. I love all kinds of animals. We went to Australia last summer and I learned all about koalas. We toured an amazing koala sanctuary and veterinary hospital called Currumbin Wildlife Sanctuary, and I even got to hold a koala named Sid! (Koalas are deceptively heavy.) In the future, I want a really modern house that is full of dogs. I am also very interested in all kinds of technology and want to learn CAD.

Where do you do STEM?
I do STEM at home, at Girl Scouts, at school, and at summer camps.

When did you first become interested in STEM?
I first became interested in STEM when I joined the Green Team at my school, which is an eco group for kids who want to change the world. We did recycling projects, helped make the indoor vertical garden for organic school lunches, and a lot of other cool environmental projects.

Why do you like doing STEM activities?
I love STEM because STEM activities pass the time when you're bored, are very fun to do, and help you learn and grow.

What is your favorite STEM project you've ever done?
Many! I especially love drawing architectural plans and designs on paper, thinking about houses and how spaces inside a house should be laid out or perhaps changed. I made up an architectural scavenger hunt with descriptions of some of the common architectural house styles in our area. It was a fun way to learn and discover (and compete with my brother!) while safely walking and driving around.

When an experiment or design doesn't turn out how you expected, how does that make you feel?
It makes me feel frustrated, and sometimes I want to quit. It's easy to think everything will go my way because I'm the one designing things or building things, but that's just not how life goes.

What other activities make you feel courageous, confident, and bold?
Sports, drawing, baking (I have a great recipe for crepes that I love to make), and traveling. I absolutely love airports and the excitement you feel when you're about to go to a new place or go somewhere that you already love.

Who do you look up to?
I really look up to my parents. My dad is an equine veterinarian, and my mom is a commercial photographer. They are smart, creative, and kind. I also look up to my

energetic and supersmart STEM teacher at school, Mrs. Corveleyn. I don't think all schools have a dedicated STEM teacher, but they should!

What are some ways girls can do STEM activities at home?
One way girls can do STEM projects at home is simply by drawing random sketches and turning them into masterpieces (or future house plans) of their own.

What advice do you have for other girls?
One piece of advice I have is to never give up—there is always a solution!

Gatee, age 10

What are some things you want us to know about you?
I like learning about the human body, and when I am at the science museum, I like to go to their paleontology lab. I love to swim and I am part of a swim team. I also enjoy drawing, face painting, calligraphy, biking, cooking, and spending time with my family.

Where do you do STEM?
I do it at home and at my local science museum.

When did you first become interested in STEM?
When I went to STEM night at my school, there was a project about how to balance erasers on one strip of paper that balanced itself on two cans. I was supposed to balance a lot of erasers, and my highest score was twenty-seven erasers.

Why do you like doing STEM activities?
There can be multiple things to do, like baking, technology, geometry. It never ends! There are multiple ways to challenge my brain. I like STEM activities because they are all so different from each other. Like, for baking I can use math to measure my ingredients, tools that use technology, and then chemistry to see their interaction at high temperature that results in yummy cookies and cupcakes.

What is your favorite STEM project you've ever done?
I like making bath bombs very much. I like how I get to use math to calculate the amount of ingredients, and I like to watch how they fizz in contact with water. I enjoy using colors and different perfumes to make them colorful and fragrant. I have flower and animal molds to make bath bombs in different shapes.

When an experiment or design doesn't turn out how you expected, how does that make you feel?
I feel a bit sad, but I also feel excited, as it gives me a chance to rearrange and rework my design. I feel very happy about my hard work once I successfully finish the process. Even though I get frustrated sometimes when I fail at first, I know that my mistakes teach me something and make me better.

What other activities make you feel courageous, confident, and bold?
Doing something that I am scared of makes me feel courageous. For example, I was really scared and nervous at my first swim meet. I almost did not want to participate in the events. But I tried to overcome my nervousness and went in the pool and actually won an event, and that made me feel courageous. Reading gives me knowledge, and that makes me feel confident. Learning a new art or activity makes me feel bold.

Who do you look up to?
I look up to my father because he is a pharmaceutical scientist, and in my backyard he built a deck and a walkway without any professional help.

What are some ways girls can do STEM activities at home?
Building a marble run using foam pieces, milk cans, cardboard boxes, paper plates, and some other household items. You can also build structures with marshmallows and toothpicks. All these projects have been done by my family and are very fun.

What advice do you have for other girls?
Never, ever give up, and never think you are not good, because everyone is special at something.

ACKNOWLEDGMENTS

This book would not have been possible without the extraordinary girls I had the privilege to interview. Thank you for entrusting me to tell your stories. Throughout this process, you have inspired me, and it is my honor to share your stories with the world. I have no doubt you will all continue to do amazing things and make a difference in our world. Please know that I'll be cheering for you every step of the way.

A huge thank you to my talented photographers: Kacey Baxter, Jenna Perfette, Chris Stanley, and Karin Belgrave. It wasn't easy working behind masks and copious amounts of hand sanitizer, but you captured the girls' personalities perfectly, allowing their unique strengths to shine on these pages.

Thank you to the entire team at Black Dog & Leventhal and Hachette Book Group for giving me this opportunity. To my editors, Lisa Tenaglia and Becky Koh, thank you for taking a chance on me, for your guidance and support, and for answering all my many questions. To my art director, Katie Benezra, thank you for helping me tell the girls' stories by turning their words and pictures into art with your skillful designs. To my managing editor, Melanie Gold, and copy editor Kelley Blewster, engineers aren't typically known for their grammar and spelling abilities, so thank you for all your diligent corrections and edits.

Thank you to my illustrator, Jim Koop. Your drawings brought these activities to life and will help guide so many girls to experiment in STEM.

I am blessed to have so much support from my parents and in-laws with this project, and in everything that I do. To my parents, Lee and Tim Pretz, thank you for always believing in me, for encouraging me to take chances, and for graciously stepping in when I need help. To my in-laws, Maureen and George Foster, thank you for the many hours of "boy watching" you happily took on so that I could have time to write.

Finally, I am eternally grateful to my family for their love. To my husband, Bryan, none of this would be possible without your encouragement and support throughout this process and over the past sixteen years. You believe in me long before I believe in myself and always push me to reach for my dreams. To my boys, Nolan and Bennett, being your mom is the best experiment of my life. Thank you for being my testers and helpers as I put together the projects in this book. Above all, thank you for teaching me how to be a better person.

IMAGE CREDITS

Kacey L. Baxter of Acorn Studios
Jenna Perfette Photography
Next Adventure Photography
Karin Belgrave Photography

ABOUT THE AUTHOR

Sarah Foster is the founder of STEM Like a Girl, a nonprofit organization formed in 2017 that aims to engage and excite elementary school girls in STEM. Sarah has a BS in chemical engineering from Bucknell University and an MS in biomedical engineering from Boston University. She worked as a research and development engineer in the biotech field before turning to educating youth, specifically girls, and their families in STEM activities. Sarah lives in Portland, Oregon, with her husband and two young boys.